Overview of Industrial Process Automation

Overview of Industrial Process Automation

KLS Sharma
Senior Professor
International Institute of Information Technology
Bangalore, India

AMSTERDAM • BOSTON • HEIDELBERG • LONDON • NEW YORK • OXFORD
PARIS • SAN DIEGO • SAN FRANCISCO • SINGAPORE • SYDNEY • TOKYO

Elsevier
32 Jamestown Road London NW1 7BY
225 Wyman Street, Waltham, MA 02451, USA

First edition 2011

Cover page background image courtesy: A&D India Magazines (www.AandD24.in)

Notices
Knowledge and best practice in this field are constantly changing. As new research and experience broaden our understanding, changes in research methods, professional practices, or medical treatment may become necessary.

Practitioners and researchers must always rely on their own experience and knowledge in evaluating and using any information, methods, compounds, or experiments described herein. In using such information or methods they should be mindful of their own safety and the safety of others, including parties for whom they have a professional responsibility.

To the fullest extent of the law, neither the Publisher nor the authors, contributors, or editors, assume any liability for any injury and/or damage to persons or property as a matter of products liability, negligence or otherwise, or from any use or operation of any methods, products, instructions, or ideas contained in the material herein.

British Library Cataloguing in Publication Data
A catalogue record for this book is available from the British Library

Library of Congress Cataloging-in-Publication Data
A catalog record for this book is available from the Library of Congress

ISBN: 978-0-323-16538-9

For information on all Elsevier publications
visit our website at elsevierdirect.com

This book has been manufactured using Print On Demand technology. Each copy is produced to order and is limited to black ink. The online version of this book will show color figures where appropriate.

This book is dedicated to ABB India, where the author learned and practiced automation for over 24 years.

Contents

Foreword

This book authored by Dr KLS Sharma, perhaps, should have been titled "Automation made easy" as he dealt with the subject matter from the very basics to its progressive development to the contemporary practices in a very illustrative manner. Dr Sharma is an academician by mind who has spent his lifetime in industry implementing automation systems in diverse industries and applications. His penchant for academics was so overwhelming that he quit a senior position in the industry to pursue his passion for research in Industrial automation and advocate its application– all across to enable Indian industry achieve global levels of productivity and thus become more competitive.

The book in indeed is a comprehensive treatise on automation covering a wide range of applications in industries ranging from discrete, continuous process to hybrid. It has dealt in detail with both software and hardware related aspects of automation applications. And interestingly, the book has also covers in detail the latest fieldbus configurations and their interfaces with the contemporary "Distributed" as well as "Open" control systems.

The world of automation is undergoing a sea change in which automation hardware is no more captive to solution-providers. The world of PLC and DCS are converging, fading away the traditional dividing lines. The application developers are set to lead the industry. All these emerging trends and much more is beautifully captured in the book.

As India, today, develops its infrastructure and industry at an unprecedented pace, it is vital that we adopt the latest and best in class technologies. We have the privilege of choice! Our large demographic size should not compel us to adopt manual or semi automatic processes which may inhibit realisation of global level of productivity and quality. In this pursuit, the scope of automation must be probed and applied. India's automation market is presently, about US$ One billion compared to China's US$ 10 Billion market. The Chinese industry is rapidly catching up with the global best in productivity and thus getting acknowledged as the "World factory."

Indian industry has to create awareness for automation benefits and also train hundred of thousand automation champions and professionals.

As a representative of Industry, I am very pleased that a book of this quality is being published which would go a long way to make the subject matter popular with the student as well as teaching community and thus give further impetus to automation applications in industry.

Ravi Uppal
President & Chief Executive Officer, Power Business
Director, Group Board
Larsen & Toubro, India

About the Author

K.L.S. Sharma graduated from Mysore University and got his masters and doctorate degrees from the Indian Institute of Technology, Delhi.

He has worked for the following organizations:

- Electronics Corporation of India Limited, Hyderabad, India
- ABB Limited, Bangalore, India
- Honeywell Technology Solutions Lab, Bangalore, India

He is currently working in the following positions:

- Senior Professor, International Institute of Information Technology, Bangalore, India
- Professor Emeritus, M S Ramaiah Institute of Technology, Bangalore, India
- Member, Editorial Advisory Board, A&D Magazine on Industrial Automation, India
- Member, Campus Connect Program, Automation Industry Association of India
- Senior Member, International Society of Automation (ISA)

Preface

During my 33-year career in the computer and automation industry and subsequently my 7 years in academic institutions, I have observed a gap between academia and industry regarding the automation domain. These observations are based on my time spent training new recruits in Indian industry and, later, on my teaching experience in Indian academic institutions. One of the ways this gap can be bridged is by introducing the basics of modern automation technology to those who are beginning careers in automation. This includes students and persons in industry who are switching to the automation domain. Prior knowledge of automation provides these beginners with a better and quicker start.

In many academic institutions, curriculum is being upgraded in instrumentation/ control engineering courses to prepare the students for careers in the automation industry. At present, the automation industry spends considerable time and money training and preparing new recruits for the job. The situation is more or less the same for persons switching to the automation domain in industry. This motivated me to write this book introducing the principles of automation in a simple and structured manner.

This book teaches beginners the basics of automation, and it is also intended as a guide to teachers and trainers who are introducing the subject. It addresses the current philosophy, technology, terminology, and practices within the automation industry using simple examples and illustrations.

The present automation system is built out of a combination of technologies, which include the following:

- Sensor and control
- Electronics
- Electrical drives
- Information (computer science and engineering)
- Communication and networking
- Embedded
- Digital signal processing
- Control engineering, and many more

Present automation technology is one of the few engineering domains that use many modern technologies. Among these, information, communication, and networking technologies have become integral part of today's automation. Basic subsystems of modern automation system are **instrumentation**, **control**, and **human interface**. In all the subsystems, the influence of various technologies is visible. By and large, the major providers of automation use similar philosophies in forming their products, systems, and solutions.

As of now, most of the information on modern automation exists in the form of technical documents prepared by automation companies. These documents are usually specific to their products, systems, solutions, and training. This knowledge has not yet been widely disseminated to the general public, and the books that are available deal with specific products and systems. Most of this industry material is somewhat difficult for beginners to understand. It is good for next-level reading after some exposure to the basics.

Automation of the industrial process calls for **industrial process automation systems**. These involve designing, developing, manufacturing, installing, commissioning, and maintaining of automation systems, which calls for the services of qualified and trained automation engineers. In addressing the basic concepts of automation, this book provides a starting point for the necessary education and training process.

Over the years, considerable advances have taken place in hardware technologies (mainly in electronics and communication). However, in computer-based automation systems, hardware interfaces have remained virtually the same—except that control has become more powerful due to availability of large memory and increased processor speeds. In other words, the memory and speed constraints present in earlier systems are not an issue today. In view of this, many aspects of hardware interface and function have been taken over by software, which does not need special interface electronics. Today, complete operation and control of industrial processes are by software-driven automation systems. Further, this software provides maximum support to the user. The user is now only required to configure and customize the automation system for a particular process. Therefore, the emphasis in this book is placed on hardware, engineering, and application programming.

This book is intended for:

- Students beginning automation careers
- Teachers of automation and related subjects
- Engineers switching to automation careers
- Trainers of automation

How to read the book:

- Follow the book from beginning to end, as its sequence is structured as a guided tour of the subject.
- Skip appendixes if you already have the background knowledge. They are provided for the sake of completeness and to create a base for easy understanding of the book.

Benefits the book provides for readers:

- Does not call for any prior knowledge of automation
- Presents a guided tour of automation
- Explains the concepts through simple illustrations and examples
- Makes further study easy
- Prepares understand technical documents in the automation industry

Based on my experience in training, teaching, and interacting with trainees and students, I have specially formulated and simplified the illustrations and discussions in this book to facilitate easy understanding.

The book is organized as follows:

Chapter 1: Why Automation?—Industrial process, Undesired behavior of process, Types and classifications of process, Unattended, manually attended, and fully automated processes, Needs and benefits of automation, Process signals

Chapter 2: Automation System Structure—Functions of automation subsystems, Instrumentation, control, and human interface, Individual roles

Chapter 3: Instrumentation Subsystem—Structure and functions, Types of instrumentation devices, Interface to control subsystem, Interfacing standards, Isolation and protection

Chapter 4: Control Subsystem—Functions and structure, Interfaces to instrumentation and human interface subsystems

Chapter 5: Human Interface Subsystem—Construction, Active display and control elements, Types of panels, Interface to control subsystem

Chapter 6: Automation Strategies—Basic strategies, Open and closed loop, Discrete, continuous, and Hybrid

Chapter 7: Programmable Control Subsystem—Processor-based subsystem, Controller, Input/output structure, Special features—communicability and self-supervisability

Chapter 8: Hardware Structure of Controller—Construction of controller, Major functional modules, Data transfer on the bus, Structure and working of functional modules, Integration

Chapter 9: Software Structure of Controller—Difference between general-purpose computing and real-time computing, Real-time operating system, Scheduling and execution of tasks, Program interrupt

Chapter 10: Programming of Controller—Programming of automation strategies, higher-level languages, IEC 61131-3 standard, Ladder logic diagram, Function block diagram

Chapter 11: Advanced Human Interface—Migration of hardwired operator panel to software-based operator station, Layout and features, Enhanced configurations, Logging station, Control desk

Chapter 12: Types of Automation Systems—Structure for localized and distributed process, Centralized system, Decentralized/distributed system, Remote/networked system, Multiple operator stations, Supervisory control and data acquisition

Chapter 13: Special Purpose Controllers—Customization of controller, Programmable logic controller, Loop controller, Controller, Remote terminal unit, PC-based controller, Programmable automation controller

Chapter 14: System Availability—Availability issues, Improvement of system availability, Cold and hot standby, Standby/redundancy for critical components

Chapter 15: Common Configurations—Configurations with operator stations, Supervisor stations, Application stations

Chapter 16: Advanced Input/Output System—Centralized I/O, Remote I/O, and Fieldbus I/O, Data communication and networking, Communication protocol

Chapter 17: Concluding Remarks—Summary, Application-wise classification of automation systems, Data handling, Future trends

Appendixes: Hardwired Control Subsystems, Processor, Hardware–Software Interfacing, Basics of Programming, Advanced Control Strategies, Power Supply System, Further Reading

I wish to emphasize that the content in this book is mainly a result of my learning, practicing, teaching, and training experience in automation areas in ABB India, where I worked for over 24 years. I would also like to mention the following organizations, where I gained valuable automation teaching and training experience:

International Institute of Information Technology, Bangalore, India (http://www.iiitb.ac.in)
National Institute of Technology Karnataka, Surathkal, India (http://www.nitk.ac.in)
Axcend Automation and Software Solutions, Bangalore, India (http://www.axcend.com)
Honeywell Technology Solutions Lab, Bangalore, India (http://www.honeywell.com)
Emerson Process Management, Mumbai, India (http://www.emerson.com)
M S Ramaiah Institute of Technology, Bangalore, India (http://www.msrit.edu)

For their help with this book, I respectfully and gratefully acknowledge the kind guidance and support of my senior colleagues

Prof. S.S. Prabhu, Senior Professor, IIIT/Bangalore; former Professor, IIT/Kanpur; and a veteran on control systems and power systems
Prof. H.N. Mahabala, former Professor, IIT/Kanpur, IIT/Chennai, IIIT/Bangalore; a veteran on information technology; and a founder of computer education in India

I also acknowledge the help of my student, Mrs. Celina Madhavan, who developed the automation program examples for the book and reviewed the manuscript.

In addition, the following professionals supported me at every stage of preparation of the manuscript with their valuable suggestions and input:

Prof. R Chandrashekar, IIIT/Bangalore, India
Mr. Hemal Desai, Emerson Process Management, Mumbai, India
Mr. Shreesha Chandra, Yokogawa, Bangalore, India

I also gratefully acknowledge the support and encouragement of Prof. S. Sadagopan, Director, IIIT/Bangalore, and Mr. Anup Wadhwa, Director, Automation Industry Association of India (AIA).

Finally, I would like to thank my wife, Mrs. Sumitra Sharma, for her kind encouragement and support through it all.

1st Mach, 2011   KLS Sharma
Bangalore, India   kls.sharma@iiitb.ac.in

Acknowledgments

I would like to gratefully acknowledge the support from the following organizations for kindly providing the information on their products, systems, and solutions for inclusion in the book:

A&D India Magazines, Pune, India	www.AandD24.in
3S-Smart Software Solutions, Kempten, Germany	www.3s-software.com
Adept Fluidyne, Pune, India	www.adeptfluidyne.com
Advantech, Seol, Korea	www.advantech.com
Ampere Technologies, Bangalore, India	www.amperetechnologies.com
Automation Technique, Pune, India	www.automationtekniks.com
Avadhoot Automation, Bangalore, India	www.avadhauto.com
Avcon, Mumbai, India	www.avconindia.com
Control Dynamics, Vadodara, India	www.mimicpanels.net
Dubas Power, Bangalore, India	www.dubaspower.com
ELCOM International, Mumbai, India	www.elcom-international.com
Emerson Process Management, Mumbai	www.emerson.com
Exide Industries, Kolkatta, India	www.exide.co.in
Fabionix, Bangalore, India	www.fabionix.co.in
General Industrial Controls, Pune, India	www.gicindia.com
Ingenious Technologies, Bangalore, India	www.aquamon.in
Integral Systems and Components, Bangalore, India	www.integralsys.com
Jai Balaji, Chennai, India	www.jaibalaji.firm.in
Jyoti, Vadodara, India	www.jyoti.com
Kalki Communication Technologies, Bangalore, India	www.kalkitech.com
Karnataka Power Corporation, Bangalore, India	www.karnatakapower.com
Kevin Technologies, Ahmadabad, India	www.kevintech.com
Levcon Group, Kolkatta, India	www.levcongroup.com
Manikant Brothers, Mumbai, India	www.accuflowmeter.com*
MECO Instruments, Mumbai, India	www.mecoinst.com
Megacraft Enterprices, Pune, India	www.megacraft.net
OEN India, Cochin, India	www.oenindia.com
Opto22, Temecula, USA	www.opto22.com
Pankaj Potentiometers, Mumbai, India	www.pankaj.com
Pepperl-Fuchs, Bangalore, India	www.pepperl-fuchs.com
Rishabh, Nashik, India	www.rishabh.co.in
Schneider Electric, Bangalore, India	www.schneider-electric.com
Secure Meters, Delhi, India	www.securetogether.com
Siemens, Mumbai, India	www.siemens.com
V-Guard, Cochin, India	www.vguard.in
Yokogawa, Bangalore, India	www.yokogawa.com

I have also consulted extensively www.wikipedia.com and many industry and education websites for the information, definitions, and images for explaining some of the concepts in automation.

KLS Sharma

*Under processing

1 Why Automation?

1.1 Introduction

Over the past few decades, the industry emphasis world-wide has been to produce quality, consistent, and cost-effective goods/services to stay in the market. Quality, consistency, and competitiveness cannot be achieved without automating the process of manufacturing the goods and of providing the service. In line with this trend, the application of automation today is omnipresent in almost all the applications, starting from deep water to the space and has gained the confidence of the world for achieving the desired results.

Over the years, automation technology has advanced along with various other technologies, such as information, communication, networking, electronics. The list below shows advances in automation technology over the past decades:

- 1940–1960: Pneumatic
- 1960–2000: Analog
- 1980–1990: Digital—proprietary
- 2000 onward: Digital—open

The current trend is to move toward an **open network of embedded systems**.[1] This chapter gives a brief introduction to automation and explains why it is necessary to automate the production of goods and providing of services to achieve the required quality, consistency, and cost for today's marketplace. In addition, there are many more complex requirements which cannot be achieved without automation.

In the subsequent discussions in the book, the term **process** refers to **industrial processes/plants** and **system** refers to **automation systems/subsystems**.

1.2 Physical Process

Physical process[2] is a series of actions, operations, changes, or functions that takes place within bringing about changes or producing an output or a result. Also, the physical process as a sequence of interdependent operations or actions which, at every stage, consume one or more inputs or resources to convert same into outputs or results to reach a known goal or the desired end result.

---

[1] Pinto, Jim. Automation Unplugged. ISA, 2004.

[2] http://www.wikipedia.org.

Overview of Industrial Process Automation. DOI: 10.1016/B978-0-12-415779-8.00001-2

Whatever we see and work within reality are all physical processes. The physical processes can be broadly divided into three categories as follows:

- Natural processes
- Self-regulated processes
- Man-made or industrial processes

1.2.1 Natural Processes

Natural processes[3] are presented by or produced by nature. The best example is a human body that, generally, does not need any external assistance to regulate its body parameters (e.g., the body temperature) irrespective of the effects of the surrounding environmental conditions. The human body maintains or regulates all its parameters. Typically, in natural processes, no abnormal behavior is present in most of the conditions.

1.2.2 Self-Regulated Processes

Self-regulated processes[4] are not natural but do not need any external assistance for their regulation. The best example is a domestic geyser in which the water level in the geyser is always maintained irrespective of its water temperature. All natural processes are self-regulated, but the reverse is not true.

1.2.3 Man-made or Industrial Processes

Man-made or industrial processes[5] are systematic series of physical, mechanical, chemical, or other kinds of operations that produce a result. They manufacture goods or provide services. These processes are not always self-regulating and may need external regulation on a continuous basis.

Typical examples of goods are any products manufactured by an industrial process, such as food, chemical, engineering". Examples of typical services include the supply of electricity, water, gas, etc., to consumers in a municipal locality.

Some simple examples of the man-made processes are discussed in the following sections.

1.2.3.1 Water Tap

The function of the simple water tap in the house is to provide water when the tap is opened. The actions involved are to open the tap and wait for the water to flow. Here, the process is the water tap, the action is turning on the tap, and the output is the water. The intention is to get the water with the desired flow. This may not always happen for several reasons, such as low or no pressure (external factors) or a clog in the pipe or tap (internal factors) resulting in either inadequate or no flow of water.

---

[3] http://www.wikipedia.org.

[4] http://www.wikipedia.org.

[5] http://www.wikipedia.org.

1.2.3.2   Electric Bulb

Similarly, the function of a simple electric bulb in the house is to provide the light when it is switched on. The actions involved are to switch the bulb on and wait for the bulb to glow. Here, the process is the electric bulb, the action is turning on the switch, and the output is the light. Once again, the intention is to get the light with the desired illumination. This may not always happen for several reasons, such as low or no voltage (external factors) or a faulty or broken filament (internal factors) resulting in either improper or no illumination.

1.2.4   Undesired Behavior

As seen in the above two examples, undesired behavior is expected under unusual conditions due to external and/or internal factors which cannot be eliminated in man-made or industrial processes. Special efforts are required to overcome or minimize the effects of these factors to ensure that the process behaves or produces the results the way we want.

   As our discussion will focus on the man-made or industrial processes and how to manage them to get the desired results, the term process will refer only to the man-made or industrial processes from now on. These processes may also be called plants or process plants.

1.3   Types of Industrial Processes

To facilitate their proper management, industrial processes are broadly divided into two categories. They are considered either localized or distributed processes, based on their nature, structure, or physical organization.

1.3.1   Localized Processes

The localized process is present in a relatively small physical area with all its sub-processes or components closely interconnected. Some simple examples of localized processes are the following:

- Water tap
- Electric bulb
- Electric motor
- Water heater
- Passenger lift
- Air-conditioner
- Traffic signal

1.3.2   Distributed Processes

Conversely, the distributed process is present in a relatively large physical area with its subprocesses or components loosely interconnected. Such a process is a **network** of many localized processes distributed over a large physical area. Some simple examples are as follows:

Several water taps (localized processes) connected to a common water supply line (water supply system) in a building. Here, the coupling or networking of the water taps is through the water pipeline supplying the water to all the taps. Each water tap is a localized process, while the group of water taps, networked through the common water supply pipe, is a distributed process.

Several electric bulbs (localized processes) connected to a common electric supply line (electric supply system) in a building. Here, the coupling or networking of the electrical bulbs is through the electricity supply cable supplying the electricity to all the bulbs. Each electrical bulb is a localized process, while the group of electrical bulbs, networked through the common electricity supply cable, is a distributed process.

The distributed process treats the entire network of localized processes as a **single entity** (not as each individual localized subprocess). Here, each localized process has some effect on the operation and performance of other networked localized processes.

Figure 1.1 explains the concepts of localized and distributed processes using the water tap and electric bulb as examples.

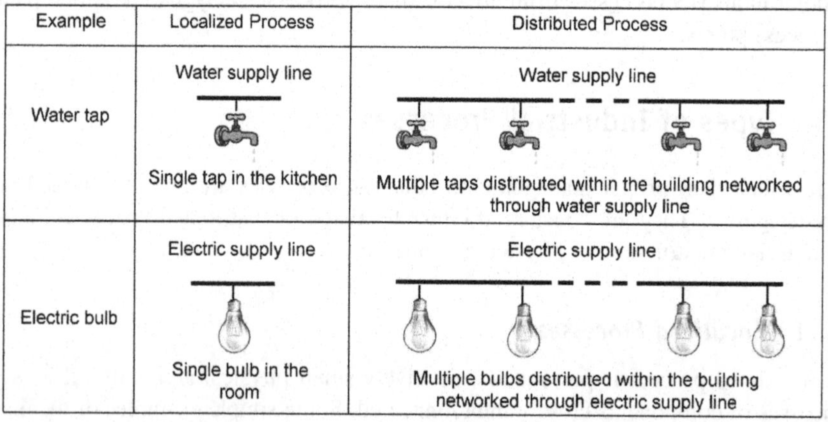

Example	Localized Process	Distributed Process
Water tap	Water supply line Single tap in the kitchen	Water supply line Multiple taps distributed within the building networked through water supply line
Electric bulb	Electric supply line Single bulb in the room	Electric supply line Multiple bulbs distributed within the building networked through electric supply line

Figure 1.1 Localized and distributed processes.

The concept of the distributed process can even be extended to include **geographically distributed processes**, such as water distribution networks, electricity distribution networks, gas distribution networks, which are normally present in a town or city or even a region. In these cases, the area covered is physically large, and the interconnection among the subprocesses or components becomes loose.

1.4   Industry Classification

The industrial processes are further broadly classified, based on their application areas, into utility industries and process industries, as described below. This is being done to address the specific technical issues present in those sectors. Not all the manufacturers of automation systems follow this classification.

1.4.1 Utility Industry

The utility industry, including electricity, water, gas, transport, is with the public service sector (normally with municipal corporations) and has both localized and distributed processes, as shown in Tables 1.1–1.3.

A water distribution system is illustrated in Figure 1.2.

An electricity distribution system is illustrated in Figure 1.3.

Table 1.1 Localized Processes

Area	Process
Water	Raw water treatment plant, sewage treatment plant, pumping station, distribution station, etc.
Electricity	Thermal power plant, hydro power plant, electrical substation, etc.
Gas	Pumping station, distribution station, etc.

Table 1.2 Geographically Distributed Process

Sector	Process
Water	Group of water pumping and distribution stations connected through water supply pipelines; for example, the city water supply and distribution system
Electricity	Group of power distribution substations connected through electricity distribution lines; for example, a city power supply and distribution system

Localized Process	Geographically Distributed Process
Main water supply line	Main water supply line
Domestic water distribution lines to houses in a single locality	Domestic water distribution lines to houses in many different localities
Single water distribution station in a locality	Multiple distribution stations networked within the distribution area (multiple localities) through water supply line

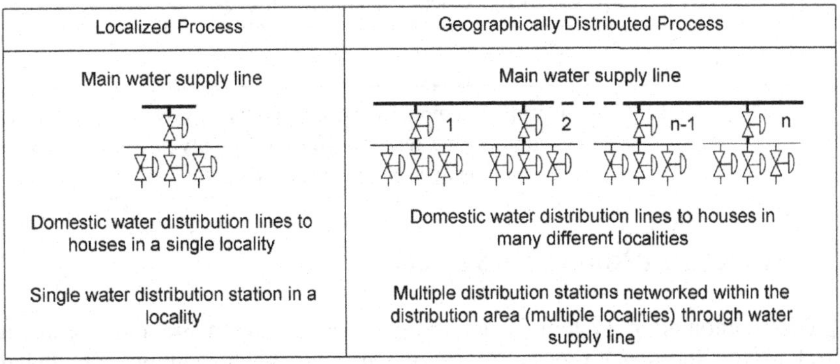

Figure 1.2 Water distribution system.

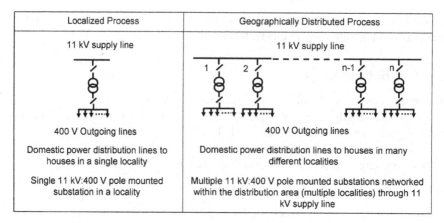

Localized Process	Geographically Distributed Process
11 kV supply line	11 kV supply line
400 V Outgoing lines	400 V Outgoing lines
Domestic power distribution lines to houses in a single locality	Domestic power distribution lines to houses in many different localities
Single 11 kV:400 V pole mounted substation in a locality	Multiple 11 kV:400 V pole mounted substations networked within the distribution area (multiple localities) through 11 kV supply line

Figure 1.3 Electricity distribution system.

1.4.2   Process Industry

The process industry is normally associated with the manufacturing or production sector, which includes chemical, metal, food, pharmaceutical. These industries have both localized and distributed processes, as laid out in Tables 1.3 and 1.4.

An oil transportation system is illustrated in Figure 1.4.

Table 1.3 Localized Processes

Sector	Process
Metal	Wire mill, plate mill, rod mill, hot strip mill, blast furnace, etc.
Chemical	Soda ash plant, demineralization plant, etc.
Oil	Refinery, pump station, block valve station, distribution terminal, etc.
Pharma	Mixing, tabletting, etc.
Food	Mixing, packing, etc.

Table 1.4 Geographically Distributed Process

Sector	Process
Oil pipeline	Group of pumping, repeater, and block valve stations connected through the oil supply pipeline; for example, petroleum product pipeline system

1.5   Process Automation System

A process automation system is an arrangement for automatic monitoring and control of the industrial process to get the desired results without any manual interventions.

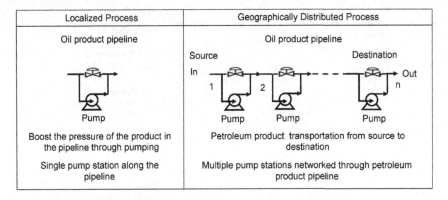

Localized Process	Geographically Distributed Process
Oil product pipeline	Oil product pipeline
Pump	Pump Pump Pump
Boost the pressure of the product in the pipeline through pumping	Petroleum product transportation from source to destination
Single pump station along the pipeline	Multiple pump stations networked through petroleum product pipeline

Figure 1.4 Oil transportation system.

Before we look into the automation, let us look at the behavior of the processes when they are unattended, manually attended, and automated. To understand the behavior of the unattended process, let us take the example of a simple water heating process, as illustrated in Figure 1.5.

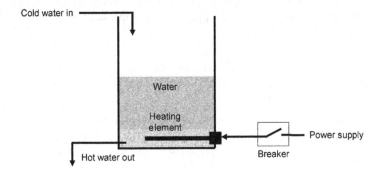

Figure 1.5 Water heating process.

1.5.1   Unattended Processes

The intention of a water heating process is to maintain the temperature of the water at a desired value. The actions involved are as follows:

- Turn on the power supply to the heating element of the water heater.
- Wait for the water to get heated.

Water continues to get heated even after crossing the desired level, if not checked. Also, no consideration is given for minimizing the effects of internal and/or external factors. This means that the heating process is totally uncontrolled and there is no guarantee that, at any point of time, we have the water with the desired temperature (see Figure 1.6).

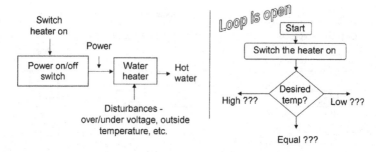

Figure 1.6 Process without automation.

There is no mechanism here to judge whether the temperature of the water is higher than, lower than, or equal to the expected level. Not only do the unattended processes not produce the desired result but they can also lead to serious consequences such as overheating of the water. Unattended processes do not produce the desired results.

1.5.2 Manually Attended Processes

To understand the behavior of the process when it is manually attended, let's take the same example of the water heater and apply the following steps to maintain the temperature of the water:

- Turn on the power supply to the water heater.
- Manually check the temperature of the water periodically.
- Manually turn off the power supply to the water heater when the temperature of the water reaches/crosses the desired level.

Following these steps, almost all the drawbacks seen in the unattended process are rectified, though with a lot of burden on the operator. The more frequently the operator checks, the better the result, but the increased work for the operator is very unproductive. The effects of both the internal and external factors are taken care of without any additional effort. Hence, in comparison to the unattended process, manually attended processes produce better, but average results. Figure 1.7 illustrates the steps involved.

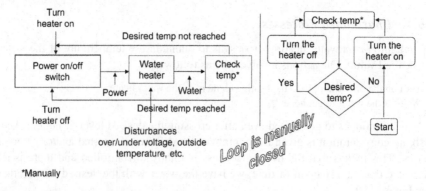

Figure 1.7 Manually attended process.

1.5.3 Automated Processes

To understand the behavior of the process when it is automated (with a temperature controller installed), let's take the same example of the water heater and apply the following steps to maintain the temperature of the water:

- Set the desired or reference temperature on the temperature controller at which the temperature of the water is to be maintained.
- Start the process by turning on the power to the heater.
- The controller continuously measures the actual temperature and keeps the heater on automatically if the actual temperature is less than the reference temperature. It turns off the heater automatically if the actual temperature is equal to or more than the reference temperature.
- Repeat the cycle.

The process continues in a loop, and the process control goes on without any manual intervention until the operator decides to stop it. The operator's job in the attended process is performed automatically by the temperature controller. Automated processes always produce the best results. Figure 1.8 illustrates the steps involved.

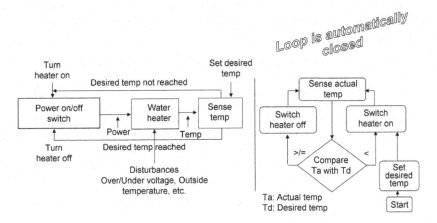

Figure 1.8 Automated process.

With the automated process, all the drawbacks seen in the unattended and manually attended process are totally rectified. The effects of both the internal and external factors are also automatically taken care of without any additional effort.

Automated processes always produce the best results. Figure 1.9 illustrates the fully automated water heating process.

1.6 Needs Met by Automation

The basic need of any process is to produce the goods or to provide the services, while conforming to environmental restrictions that maintain the following:

- Consistency
- Quality
- Cost-effectiveness

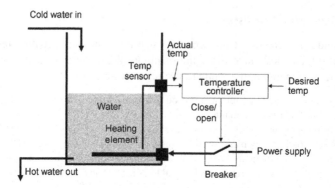

Figure 1.9 Automated water heating process.

We have seen that only the automated process (without any human intervention) produces excellent results, while the unattended or manually attended processes generally do not always produce the desired results. While the foremost intention of any process is to meet the desired result, there are other important considerations to be met. Automation also manages the following:

• Hazardous processes (nuclear, high voltage, toxic, etc.) where human intervention is quite dangerous and is not desirable.
• Repetitive processes (traffic management, etc.) where continuous, repetitive manual operations can lead to failure due to human fatigue.
• Sequential startup and shutdown of plants (power plants, chemical plants, etc.) while satisfying certain conditions at each step where sequential manual operation, with safety and/or other conditions to be satisfied, is highly time-consuming and prone to errors.
• Complex processes (decisions based on heavy computing such as aircraft tracking and guiding) where manual computing and decision making within a short time is next to impossible.

There are many more complex processes with needs that can be met most efficiently through automation.

1.7 Benefits of Automation

In addition to attaining the desired quality, consistency, and cost-effectiveness of the products and services, the benefits of automation are as follows:

• Reduction of losses by the fastest possible restoration or restarting of the plant after the following situations:
  · Breakdown of the plant
  · Reconfiguring of the plant to adjust to new requirements
• In practice, the reduction of production loss can also be described as a **reduction in unproductive time** for the plant. This way, the production of goods or services is increased, contributing to higher revenue and profit. This is also known as an increase in the overall availability of the plant.

- Optimization of resources through substantial reduction of dependence on highly skilled manpower
- Higher safety for personnel and equipment, as they do not come in contact with the working of the plant
- Higher security and reliability in the operation of the plant
- Faster response and result, as there is no human intervention required

All of these benefits lead to lesser overall operational costs and higher operational efficiency.

1.8   Automation Steps

As discussed earlier, a typical automation system exccutes the automation steps sequentially and cyclically, as shown in Figure 1.10. This serves to continuously effect corrections to the process operation, and it consistently produces the desired results.

The automation steps are explained as follows.

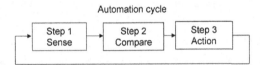

Figure 1.10 Typical automation steps and cycle.

1.8.1   Step 1: Information Acquisition

Information acquisition step observes the behavior of the process by sensing or measuring process parameters of interest. These parameters are called process inputs.

1.8.2   Step 2: Information Analysis and Decision Making

The information analysis and decision making step analyse the behavior of the process by comparing the acquired information with the desired result. Then a decision is made about the new directives or commands that would be required to effect any corrections.

1.8.3   Step 3: Control Execution

Control execution step actually controls the behavior of the process by sending any new directives or commands into the process to effect the corrections. These directives are called command outputs.

These steps are illustrated in Figure 1.11 using the example of a water heater.

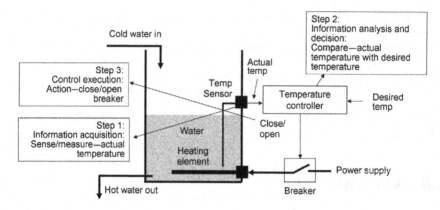

Figure 1.11 Automation steps and cycle in water heating process.

1.9 Process Signals

A process signal[6] is a fluctuating or varying physical quantity (such as electrical, mechanical, chemical, thermal) whose variations represent the coded or embedded information of the process parameters.

1.9.1 Classifications

In practice, the process signals exist in one of the physical forms shown in Table 1.5.

Normally, all the discussions on discrete or digital signals are equally valid for fluctuating or pulse signals, as they are variations of discrete or digital signals.

1.9.2 Input and Output Signals

The process signals are further divided into input and output signals (I/O signals). The signals that are sent by the process and received by the automation system (to understand the process behavior) are called input signals. The signals that are sent by the automation system (to change the process behavior) and received by the process are called output signals.

The input and output signals are always defined with reference to the automation system.

The input and output signals can be in any form: discrete, continuous, or fluctuating. The form depends on which parameters of the process are to be monitored and which parameters of the process are to be controlled.

1.9.2.1 Input Signals

Typical examples of input signals (measurements) are as follows:

- *Discrete*: Breaker status (on or off), control valve status (open or closed).
- *Continuous*: Temperature value, pressure value, level value, flow value.
- *Fluctuating*: Energy consumption readings, water consumption readings.

[6] http://www.wikipedia.org.

Table 1.5 Process Signals

Signal	Definition	Amplitude vs. Time
Discrete (digital)	States changing discretely with time. The information here is the current state.	
Continuous (analog)	Amplitude changing continuously with time. The information here is the current value.	
Fluctuating (pulse)	A variation of a discrete (digital) signal, changing its state more frequently. The information here is the current fluctuation. The fluctuating/pulse signals may be periodic or not periodic.	

1.9.2.2 Output Signals

Typical examples of output signals (commands) are as follows:

- *Discrete*: Open/close breaker, open/close control valve, start/stop motor.
- *Continuous*: Vary control valve opening/closing, Vary voltage output of the drive.
- *Fluctuating*: Stepper motor control.

Figure 1.12 illustrates various types of input and output signals in the water heating process, including level control and water and power consumption reading.

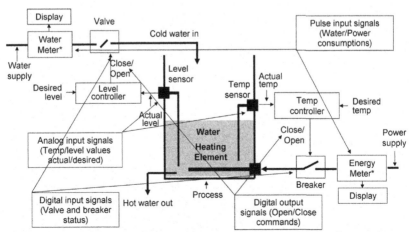

*Water and energy meters generate and output pulses proportional to the consumption of water and power

Figure 1.12 Process signals in water heating process.

1.10   Summary

In this introductory chapter, we have covered the basics of a process. We examined different types of processes, process classifications based on their nature and application, steps in automation, process needs to be met by automation, benefits of automation, and process signals and their classifications. These introductory concepts form the basis for further detailed discussions in the following chapters.

2 Automation System Structure

2.1 Introduction

In Chapter 1, we briefly discussed the automation cycle and its steps. In this chapter, those concepts are further elaborated to define the structure of the overall automation system and the functions of its subsystems.

2.2 Subsystems

The automation system is broadly divided into three subsystems as follows:

- Instrumentation subsystem
- Control subsystem
- Human interface subsystem

The interconnections and information flow among these subsystems are illustrated in Figure 2.1.

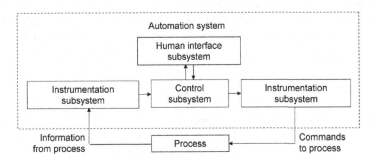

Figure 2.1 Basic structure of automation system.

2.2.1 Instrumentation Subsystem

An instrumentation subsystem acquires information on the behavior of the process (through measurement of the process parameters) and sends this to the control subsystem in an acceptable form. In the other direction, the instrumentation subsystem sends the information to the process in an acceptable form to change the behavior of the process (control of process parameters).

Overview of Industrial Process Automation. DOI: 10.1016/B978-0-12-415779-8.00002-4

2.2.2 Human Interface Subsystem

The human interface subsystem allows the operator to manually interact with the process. The operator may observe and monitor what is happening inside the process or issue manual commands, if required, to force a change in the process behavior.

2.2.3 Control Subsystem

Control subsystem is the heart of automation system and performs the following functions:

With instrumentation subsystem:

- Acquires information continuously via instrumentation on the behavior of the process.
- Compares the received information with the desired behavior of the process.
- Decides on actions on whether or not to issue commands for correcting the behavior of the process.

With the human interface subsystem:

- Acquires the information continuously from the human interface subsystem.
- Routes the received information to the process for manual control via instrumentation.
- Collects the information from the process and routes it to the human interface subsystem for display.

To sum up, the control subsystem manages the information flow to and from the instrumentation subsystem for process monitoring and control, and to and from the human interface subsystem for manual interaction with the process.

For a review of functions of the various subsystems, revisit the fully automated water heating process discussed in Chapter 1.

Figure 2.2 illustrates the fully automated water heating process with level and temperature controls with facility to measure water and power consumption.

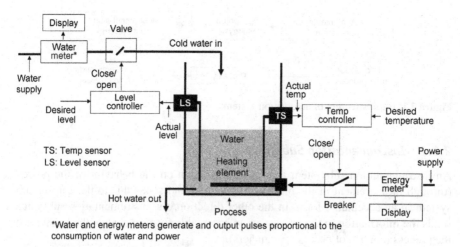

*Water and energy meters generate and output pulses proportional to the consumption of water and power

Figure 2.2 Automation system for water heating process.

The functions continuously performed by the automation system, after setting the desired values for temperature and level, are as follows:

- The temperature controller checks the actual water temperature to find out whether it is higher than or equal to the desired temperature. The controller then opens or closes the breaker accordingly.
- The level controller checks the actual water level to find out whether it is higher than or equal to the desired level. The controller then opens or closes the valve accordingly.

2.3 Instrumentation Subsystem

The instrumentation subsystem[1] is the branch of engineering that deals with the measurement and control of process parameters. An instrument is a device that measures the physical variable and/or manipulates it to produce an output in an acceptable form for further processing in the next stage device.

Instrumentation devices are not normally intelligent and are not capable of making decisions (there are exceptions discussed in Chapter 16).

As illustrated in Figure 2.3, the instrumentation subsystem is the interface between the physical process and the control subsystem for both the measurement of information and the transfer of control commands to the process.

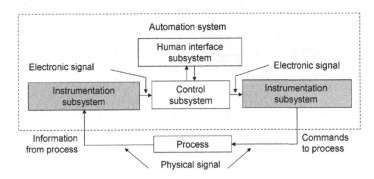

Figure 2.3 Instrumentation subsystem.

2.3.1 Measurement of Information

The present day control systems, which are electronics-based, understand and process signals only in the electronic form. Hence, in the measurement of information from the process, instrumentation devices need to convert the physical process signals (electrical, mechanical, chemical, etc.) into their equivalent electronic representations so that the control subsystem can accept and process them. **This conversion into electronic units is carried out without any loss of information.**

[1] http://www.wikipedia.org

Figures 2.4–2.6 illustrate the instrumentation devices employed for measurement in a fully automated water heating system.

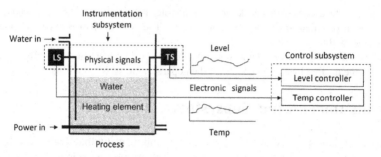

Figure 2.4 Measurement of water level and temperature.

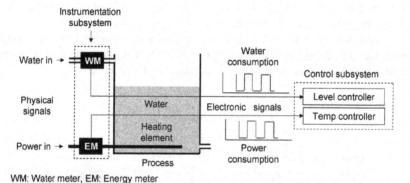

Figure 2.5 Measurement of water and power consumption.

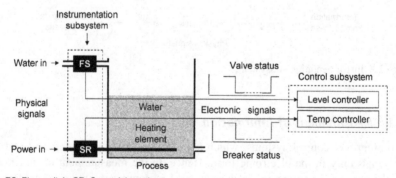

Figure 2.6 Measurement of the status of valve and breaker.

Level and temperature, being physical signals, are measured, converted into their electronic equivalents, and sent to the control subsystem by temperature and level instrumentation devices (analog inputs). The temperature sensor and level sensor measure the actual temperature and actual level.

Water and power consumption, in their physical form, are measured, converted into their electronic equivalents, and sent to the control subsystem by water and energy meters (pulse inputs). The water meter measures the consumption of water, while the energy meter measures the consumption of power.

The valve and breaker status are measured **indirectly** as presence or absence of water flow in the pipe and presence or absence of power flow in the line (indicating valve open/closed and breaker on/off). These status measurements are converted into their electronic equivalents and sent to the control subsystem by a flow switch and a supervision relay or voltage switch (digital inputs). The flow switch detects the presence/absence of water flow in the pipe, while the supervision relay detects the presence/absence of electrical current in the heater.

2.3.2   Transfer of Control Command

Since the process only understands physical signals, the electronic control signals generated by the control subsystem are converted into their equivalent physical representations (electrical, mechanical, chemical, etc.) in an acceptable form for the process to receive the control commands from the control subsystem. **Here also, the conversion or transformation is carried out without any loss of information**.

Figure 2.7 illustrates the role of instrumentation devices for control command transfer in the fully automated water heating system.

Control commands (to open/close the valve and open/close the breaker) generated by the control subsystem in electronic form are converted into their physical equivalents (electronic to mechanical) and sent to the process by a solenoid control valve and electromechanical contactor (digital outputs). The solenoid valve opens or closes the valve, while the control relay allows or disallows the power to the heater.

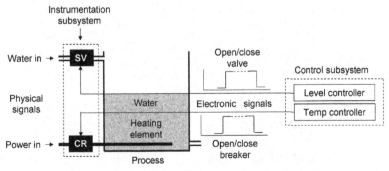

SV: ON-Off Solenoid valve, CR: On-Off Control relay

Figure 2.7 Control command transfer.

2.4   Human Interface Subsystem

The human interface subsystem, also called human–machine interface (HMI) or man–machine interface (MMI) or human–system interface (HSI), is the means by which the users or operators manually interact with the process.

As illustrated in Figure 2.8, the human interface subsystem is the interface between the physical processes and the operator, via the control subsystem, for both direct monitoring of the process parameters and to effect manual control of the process.

There is no need for any interface devices between the control subsystem and human interface subsystem (unlike the need for the instrumentation subsystem between the process and control subsystem). The signals between the control and human interface subsystems are electronic in both directions and therefore compatible.

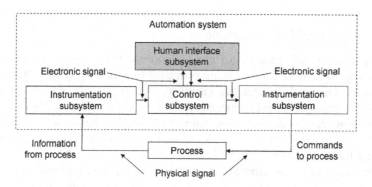

Figure 2.8 Human interface subsystem.

Some typical examples of human interface subsystems we come across in our daily life are illustrated in Figure 2.9.

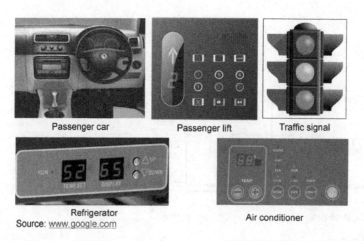

Source: www.google.com

Figure 2.9 Examples of human interfaces.

2.4.1 Manual Display and Monitoring

A manual display and monitoring provision is for directly observing the process variables of interest manually whenever there is a need. In other words, the operators can observe exactly what is happening inside the process through visual display of the parameters of interest.

2.4.2 Manual Control

The manual control provision is for manually manipulating or controlling the process parameters of interest whenever there is a need. This action overrides the functions performed automatically by the control subsystem. The provision for setting the reference values for limit checking is part of the human interface subsystem.

Figure 2.10 illustrates the implementation of the hardware-based human interface or operator panel for interaction with the water heater process. The operator panel allows the operators to manually do the following:

- Observe the actual values of temperature and level, actual consumption of water and power, and actual status of valve and breaker.
- Set the reference values for temperature and level.
- Control (close or open) the valve and breaker.

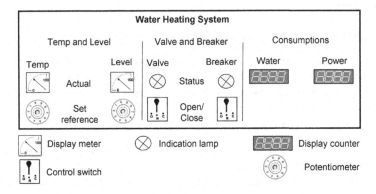

Figure 2.10 Operator panel for water heating system.

2.5 Control Subsystem

The control subsystem[2] is a mechanism or a device for manipulating the output of a process and for managing, commanding, directing, or regulating the behavior of the process to achieve the desired result.

Control subsystems are intelligent devices and are capable of making decisions.

[2] http://www.wikipedia.org

As illustrated in Figure 2.11, the control subsystem is the heart of the automation system, and it manages the instrumentation and human interface subsystems as well as its own automatic control functions.

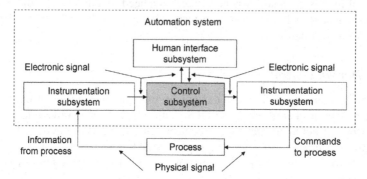

Figure 2.11 Control subsystem.

The control subsystem performs the following functional steps in sequence.

2.5.1 Information Acquisition

In this step, the control subsystem acquires the process information through the instrumentation subsystem.

2.5.2 Information Analysis and Decision Making

In this step, the control subsystem analyzes the acquired information by comparing it with the preset reference values for deviations and decides whether to effect changes in the behavior of the process or not through commands.

2.5.3 Control Execution

In this step, the control commands are finally executed (sending them to the process through the instrumentation devices).

These steps can be explained pictorially as a cyclic operation or automation cycle, as illustrated in Figure 2.12.

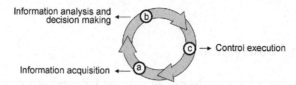

Figure 2.12 Typical automation cycle.

Figures 2.13 and 2.14 illustrate the input/output relationship of the control subsystems to both the instrumentation and human interface subsystems for level and temperature control in the water heating process.

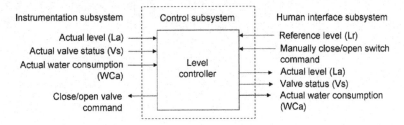

Figure 2.13 Input/output relationship in level controller.

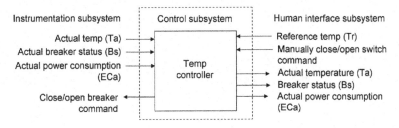

Figure 2.14 Input/output relationship in temperature controller.

The temperature and level controllers, built with appropriate automation strategies, operate on the inputs from the process and the human interface subsystem, and they produce the control command outputs to drive the process and human interface subsystem for display.

The question of how the automation strategy is embedded in the control subsystem and how it works is discussed in subsequent chapters.

2.6 Summary

In this chapter, we discussed the overall structure of the automation system as well as its functional subsystems and interconnections. These functional subsystems are discussed in greater detail in the following chapters.

3 Instrumentation Subsystem

3.1 Introduction

In the previous chapter, we discussed the basic subsystems of the automation system. Of these subsystems, the instrumentation subsystem is the most important. Without the instrumentation subsystem, the automation would not function. In this chapter, the instrumentation devices that comprise this subsystem are discussed in detail.

Instrumentation[1] is the branch of engineering that deals with the measurement and control of physical process parameters. As explained in Chapter 2, the instrumentation devices provide an interface between the process and the control subsystem to facilitate the following:

*   *Information acquisition*: It is the conversion of the physical signal generated by the process into an equivalent and compatible electronic signal (without any loss of information) acceptable to the control subsystem.
*   *Control execution*: It is the conversion of the electronic control signal generated by the control subsystem into an equivalent and compatible physical signal (without any loss of information) acceptable to the process.

3.2 Structure

The instrumentation devices have different structures for different input and output signals; they are different for continuous/analog, discrete/digital, and fluctuating/pulse signals.

3.2.1 Continuous/Analog Instrumentation Devices

Figure 3.1 illustrates the functional components and the general structure of continuous/analog instrumentation devices, for both input and output.

3.2.1.1 Information Acquisition

In Figure 3.2, the structure and the input/output relationship of an analog input instrumentation device is illustrated.

The analog input instrumentation device processes the received physical signal generated by the process with the following functional components:

---

[1] http://www.wikipedia.org

Overview of Industrial Process Automation. DOI: 10.1016/B978-0-12-415779-8.00003-6

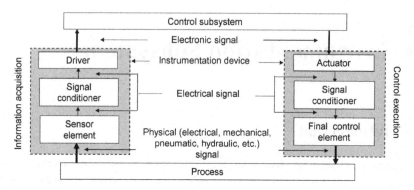

Figure 3.1 Analog input and output instrumentation device.

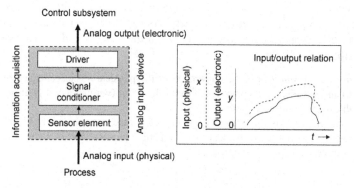

Figure 3.2 Analog input instrumentation device—input/output relationship.

- *Sensor*: It converts an analog or continuous physical signal into its electrical equivalent signal through the transduction process.
- *Signal conditioner*: Used in the intermediate stage, the signal conditioner prepares or manipulates (isolating, compensating, linearizing, amplifying, filtering, offset correcting, etc.) the weak and noisy sensor output to meet the requirements of the next stage for further processing.
- *Driver*: Strengthens the signal received from the signal conditioner to an appropriate form and drives it for transmission to the control subsystem.

The instrumentation device is called a **transducer** if its output is suitable for transmission (in voltage or current form) over a relatively short distance. The device is called a **transmitter** if the output is suitable for transmission (in current form) over a relatively long distance.

Figure 3.3 illustrates the functions of the components of the analog input instrumentation device.

Figure 3.4 illustrates industry examples of four commonly used analog input instrumentation devices in process industry (temperature transmitter, flow transmitter, level transmitter, and differential pressure transmitter). These devices receive the physical signals (temperature, flow, level, differential pressure) and generate corresponding electronic output in current form suitable for the control subsystems (see Section 3.4).

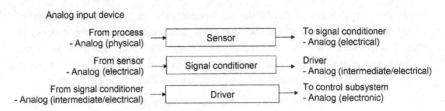

Figure 3.3 Analog input instrumentation device—components.

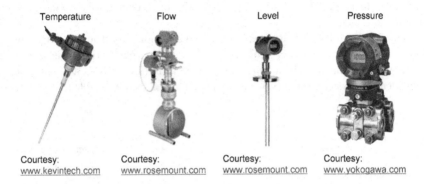

Figure 3.4 Analog input instrumentation devices—examples.

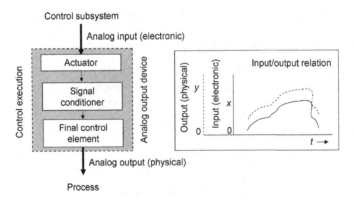

Figure 3.5 Analog output instrumentation device—input/output relationship.

As seen in Figure 3.4, these devices also support local display of the process parameters in engineering units to help field maintenance personnel.

3.2.1.2 Control Execution

Figure 3.5 illustrates the structure and input/output relationship of an analog output instrumentation device for the transmission of continuous or analog signals (commands) for the process for control execution.

The analog output instrumentation device processes the electronic control signal generated by the control system to produce the final value acceptable to the process. The associated components are the actuator, signal conditioner, and final control element. While the signal conditioner has the same function as in the analog input instrumentation device, the functions of the actuator and final control element are given below:

- *Actuator*: Transforms the command output electronic signal from the control system into a signal suitable for the signal conditioner.
- *Final control element*: Performs the final control action in the process.

Figure 3.6 illustrates the functions of all the components of an analog output instrumentation device.

Figure 3.7 illustrates industry examples of two commonly used analog output instrumentation devices (variable control valve and variable speed drive). These devices receive the electronic signals in current or voltage form (see Section 3.4) and generate corresponding physical output (valve opening and motor speed) suitable for the process.

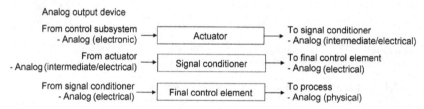

Figure 3.6 Analog output instrumentation device—components.

Figure 3.7 Analog output instrumentation devices—examples.

3.2.2   Discrete/Digital Instrumentation Devices

Figure 3.8 illustrates the functional components and the general structure of a discrete/digital instrumentation device (both input and output).

3.2.2.1   Information Acquisition

Figure 3.9 illustrates the structure and input/output relationship of a digital input instrumentation device used for information acquisition of discrete or digital signals from the process.

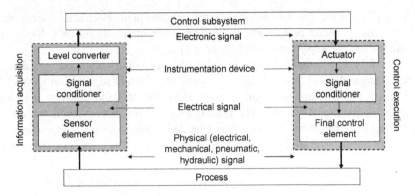

Figure 3.8 Digital input and output instrumentation device.

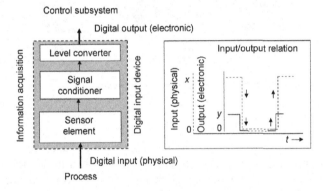

Figure 3.9 Digital input instrumentation device—input/output relationship.

The digital input instrumentation device processes the physical signal generated by the process to make it suitable for the control subsystem. The associated components are sensor, signal conditioner, and level converter. While the roles of sensor and signal conditioner remain the same, the role of *level converter* is to change the level of the electrical input (presence or absence) to produce an output (presence or absence) that meets the control subsystem requirements for processing.

Figure 3.10 illustrates the functions of the components of a digital input instrumentation device.

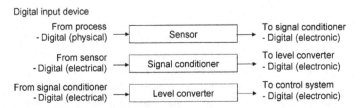

Figure 3.10 Digital input instrumentation device—components.

These devices receive the physical signals (mechanical, electrical, optical, voltage, etc.) and generate corresponding electronic output in current or voltage form suitable for control subsystems (see Section 3.4).

Figure 3.11 illustrates industry examples of three commonly used digital input instrumentation devices (limit switch, proximity switch, and supervision relay). The limit switch receives the mechanical input signal (on/off) and converts it into an electronic output signal (on/off). The inductive optical proximity switch, through its electromagnetic/optical field, senses the target and converts it into an electronic output signal (on/off). The supervision relay picks up voltage and gives a yes/no signal.

Limit switch   Proximity switch   Supervision relay

Courtesy: Courtesy: Courtesy:
www.jaibalaji.firm.in www.pepperl-fuchs.com   www.jyoti.com

Figure 3.11 Digital input instrumentation devices—examples.

3.2.2.2 Control Execution

Figure 3.12 illustrates the structure and the input/output relationship of a discrete/digital output instrumentation device (digital control device) for the transmission of discrete or digital signals (commands) for the process or control execution.

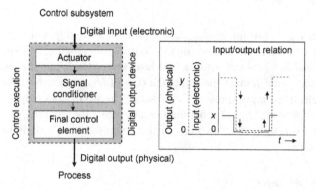

Figure 3.12 Digital output instrumentation device—input/output relationship.

The digital output instrumentation device processes the electronic signal generated by the control subsystem, making it suitable for the process. The associated components are the actuator, signal conditioner, and the final control element.

Figure 3.13 illustrates the functions of the components of a digital output instrumentation device.

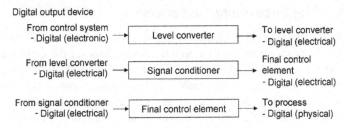

Figure 3.13 Digital output instrumentation device—components.

Figure 3.14 illustrates industry examples of two commonly used digital output instrumentation devices in the industry (on/off solenoid valve and control relay). These devices receive the electronic signals in current or voltage form (see Section 3.4) from the control subsystem and generate corresponding physical output (open/close the flow and open/close the breaker) suitable for the process.

Figure 3.14 Digital output instrumentation devices—examples.

3.2.3   Fluctuating/Pulse Signals

The discussions of digital instrumentation devices apply to fluctuating or pulse signals too, as they are variations of digital signals. Pulses are received from the process in physical form by pulse input instrumentation device, and they are sent to the control subsystem after conversion into compatible electronic form. In the reverse direction, the pulses in electronic form are sent by the control subsystem, received by the pulse output instrumentation device, and sent to the process after conversion into compatible physical form. These operations are similar to those of digital input/output instrumentation devices. The only difference is that the input and output generally changes its state more frequently. Digital input/output instrumentation devices can also provide the necessary interface for pulse input and output.

3.3   Special Instrumentation Devices

Some input instrumentation devices have special interfacing functions. These special instrumentation devices receive analog input from the process, manipulate it, and produce an output in different form.

3.3.1   Switching Instrumentation Devices

A switching instrumentation device is an input device that receives continuous or analog input and produces discrete or digital output whenever the input exceeds the **locally preset** reference value. This device is a simple **stand-alone automation system** with its own instrumentation (analog measurement), control (comparing and decision making), and human interface (setting of reference value) all integrated in one. This device can also be called a switch or stand-alone limit-checking device.

Figure 3.15 illustrates the structure and the input/output relationship of the **switch** (analog input/digital output device), which is used for information acquisition.

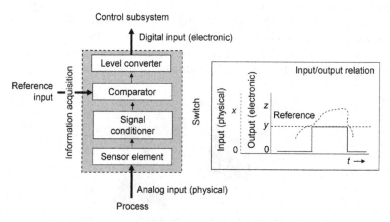

Figure 3.15 Switching instrumentation device—input/output relationship.

Figure 3.16 illustrates the functions of the components of the switching instrumentation device. While the functions of the level converter, signal conditioner, and sensor element are the same, the **comparator** produces a digital output (change of state) whenever the input analog signal violates the locally preset limit in the device and on returning to normal.

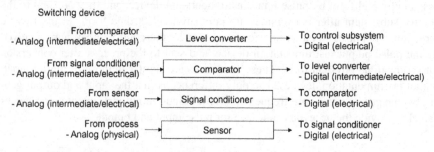

Figure 3.16 Switching instrumentation device—components.

Figure 3.17 illustrates industry examples of three commonly used instrumentation switches (flow switch, level switch, voltage switch). These devices, with provisions for setting the limit values, receive the physical signals (flow, level, and voltage) and generate corresponding electronic output in current or voltage form suitable for control subsystems (see Section 3.4). These devices are less expensive compared to the transmitters.

Figure 3.17 Switching instrumentation devices—examples.

3.3.2 Integrating Instrumentation Devices

The integrating instrumentation device is another type of special input device which receives continuous or analog input and produces fluctuating or pulse output by integrating the input over time. It converts the integrated value into proportional sequence pulses and sends the pulses to the control system in the same order in a compatible electronic form.

Figure 3.18 illustrates the structure and the input/output relationship of the **integrator** (analog input/pulse output device), which is used for information acquisition.

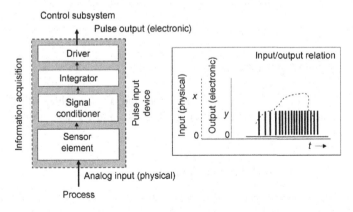

Figure 3.18 Integrating instrumentation device—input/output relationship.

Figure 3.19 illustrates the functions of the components of an integrating instrumentation device. While the functions of the sensor, signal conditioner, and level converter are the same, the integrator produces a sequence of pulse output corresponding to the integrated value of the input.

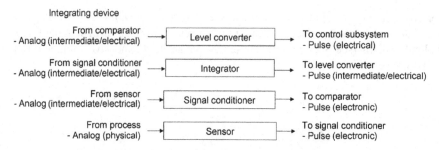

Figure 3.19 Integrating instrumentation device—components.

Figure 3.20 illustrates industry examples of two commonly used integrating instrumentation devices, which compute consumption (water meter and energy meter). These devices, with provisions for local displays in engineering units, receive the physical signals (water flow and voltage/current) and generate corresponding electronic pulse output in suitable current or voltage form for the control subsystems (see Section 3.4).

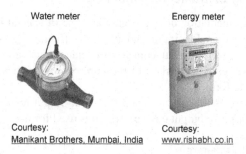

Figure 3.20 Integrating instrumentation devices—examples.

3.4   Interfacing Standards

To make the instrumentation devices compatible and interoperable with the control systems that are manufactured by different vendors, industry standards are defined for interfacing.

3.4.1   Analog Input and Output Devices

The interfacing standards for analog input and output devices are illustrated in Figure 3.21. The most common standard is **4–20 mA**.

3.4.2   Digital Input and Output Devices

The interfacing standards for digital input and output devices are illustrated in Figure 3.22. The most common standard is **24 V DC**.

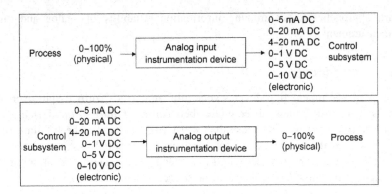

Figure 3.21 Analog instrumentation devices—interfacing standards.

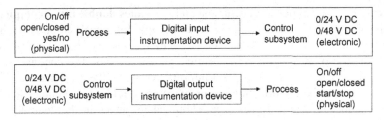

Figure 3.22 Digital instrumentation devices—interfacing standards.

3.4.3   Switching and Integrating Devices

The interfacing standards for switching and integrating devices are illustrated in Figure 3.23. Here, both the devices are input devices, and they follow the standards for digital input devices.

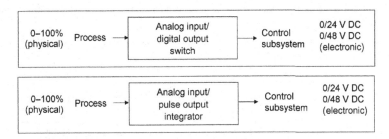

Figure 3.23 Switching/integrating instrumentation devices—interfacing standards.

3.5   Information Reliability

Even though the standards specify many possibilities, to increase the reliability of information acquisition, some special techniques are employed. Following are the standard approaches to maximizing information reliability of analog and digital input measurement.

3.5.1 Analog Inputs

Analog input instrumentation devices generally support 4–20 mA output over other standards. This arrangement differentiates between the *failure of the device* and *no input to the device*. Any output less than 4 mA is considered to be a failure of the device. However, these devices call for auxiliary power supply support to function, while some are self-powered, deriving the power for their internal functioning from the physical signal input (e.g., electrical transducers).

3.5.2 Digital Inputs

In the case of digital input, the instrumentation generates **double bit** or **complementary** signals (00 or 01 or 10 or 11) instead of **single bit** signals (0 or 1) for inputting to the control subsystem. The control system acquires the complementary input through the use of two digital input instrumentation devices. The combinations 01 and 10 are accepted and used as good information, while the combinations 00 and 11 are rejected as bad information (complementary errors). This doubles the input instrumentation devices and consumes additional resources in the control subsystem. *This feature is normally used in switchgears (circuit breakers, isolators, etc.) in electrical industry.*

3.6   Isolation and Protection

Generally, the control center is not located close to the process, but instrumentation devices are installed close to the process equipment in the plant for technical reasons. Field cables (signal/control) are laid to send/receive electronic signals between the control center and the process.

Typical interconnections among the process, instrumentation, and control subsystem are illustrated in Figure 3.24.

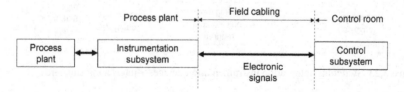

Figure 3.24 Automation system layout in process plant.

Figure 3.25 shows a typical location of an instrumentation device in the process plant.

Courtesy: www.emerson.com

Figure 3.25 Instrumentation device in process plant.

The presence of interconnecting field cables raises a few safety issues for process plants, control systems, and operating personnel. This section discusses safety issues (isolation and protection) and how to handle them. The typical examples are:

- Electrical fault at in the input signal affecting the output signal and vice versa.
- Explosion due to a spark in the plant where potentially hazardous gas/vapor is present.

3.6.1 Isolation

The fault in input signal can be caused by grounding, high voltage, short circuit, faulty devices, etc. In order to avoid the effects of faults getting transferred to control center and/or instrumentation devices, there is a need for suitable isolating devices between the control center and the instrumentation devices. This arrangement protects control subsystem, operating personnel, and instrumentation devices.

3.6.2 Protection

Incidence of ignitable spark due to high energy (beyond the acceptable limit) travel from control center into the field area through the field signal/control cables. This situation is very common in petrochemical complexes. The spark can lead to severe fire hazards and even explosions, as the environment may be charged with ignitable gaseous vapors. This calls for protection devices.

3.6.3 Solutions

Figure 3.26 illustrates the preferred solution for overcoming these safety problems by using appropriate isolation and protection units. These are called **intrinsic safety isolated barriers** or **isolated barriers**.

The isolated barrier is an electronic device for the following:

- Isolating the signal in either direction
- Protecting the equipment by limiting current, voltage, and total energy delivered from safe area to hazardous area

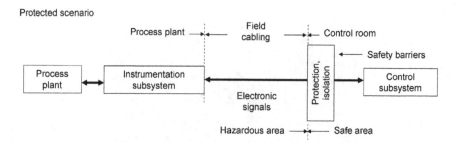

Figure 3.26 Application of barriers.

These circuits are not generally integral parts of the instrumentation subsystem, and they are installed in the control center. Figure 3.27 illustrates a typical circuit schematic of an isolated safety barrier.

In Fig. 3.27, the roles played by the various electronic components are as follows:

- *Resistor*: limits current.
- *Fuse*: protects zener diodes on over-current.
- *Zener diode*: clips voltage on over-voltage.
- *DC/DC Transformer*: provides galvanic isolation.

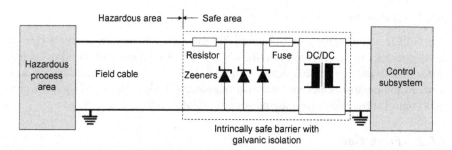

Figure 3.27 Safety barrier with isolation.

In essence, the safety barriers limit the energy transferred from the safe area to the hazardous area.

Figure 3.28 illustrates an industry example of intrinsically safe barriers stacked and mounted on a DIN rail.

Courtesy: www.pepperl-fuchs.com

Figure 3.28 Intrinsically safe barrier—example.

3.7 Summary

In this chapter, we discussed in detail the construction of various types of instrumentation devices—analog, digital, pulse, and special. Among these, we distinguished between input and output devices with industry examples. We also discussed the interfacing standards to make the instrumentation devices manufactured by different vendors compatible and inter-operable with the control subsystems from different vendors. Finally, this chapter also covered the safety issues and the need for isolating and protecting the equipment and operating personnel from hazards.

4 Control Subsystem

4.1 Introduction

The control subsystem is the heart of automation systems and coordinates and controls the functions of other subsystems as described in Chapter 2. A control subsystem[1] is a device or a set of devices to manage, command, direct, or regulate the behavior of the process. The control subsystem is an essential mechanism for manipulating the output of a specific process to achieve the desired result. This subsystem performs the following functions:

- **Information acquisition** from the process via instrumentation subsystem (process inputs) and from the human interface subsystem (manual inputs)
- Information processing, monitoring, analysis, and decision making for corrections to the behavior of the process, if required
- Generation and transmission of control commands to the process (control execution) via instrumentation subsystem and driving process parameters of interest for display on the human interface subsystem

4.2 Structure

In addition to its own functions or the **automation strategy** (information processing and decision making), the control subsystem has appropriate interfaces for interaction for information exchange with the instrumentation and human interface subsystems. In other words, the control subsystem is divided into automation strategy (responsible for managing automatic control) and interfaces for information exchange with instrumentation and human interface subsystems.

The general structure of the control subsystem is illustrated in Figure 4.1.

The interface shown in this figure is an arrangement to do the following:

- Convert the information received from the process (via instrumentation) and human interface subsystems in electronic form to a form acceptable to the automation strategy in the control subsystem.
- Convert the information generated by the automation strategy in the control subsystem into a form acceptable to the process (via instrumentation) and human interface subsystems.

These conversions and transfers, once again, should happen without any loss of information.

The automation strategy is responsible for managing the previously discussed functions through the interfaces and for performing automatic control of the process.

[1] www.wikipedia.org

Overview of Industrial Process Automation. DOI: 10.1016/B978-0-12-415779-8.00004-8

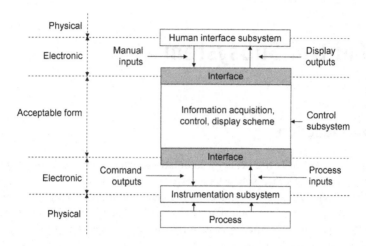

Figure 4.1 Control subsystem—structure.

4.3 Interfacing

The following sections explain the interfacing between the control subsystem and the instrumentation and human interface subsystems.

4.3.1 General

To facilitate the information transfer to and from the control subsystem, the interface is structured on inputs/outputs basis, as illustrated in Figure 4.2. Further, these inputs and the outputs can be either analog or digital or pulse. This structure leads to two main categories as inputs and outputs and, within each, three sub-categories (digital, analog, and pulse). To summarize, the interfaces consist of the following inputs and outputs:

- Digital inputs (DI), analog inputs (AI), and pulse inputs (PI)
- Digital outputs (DO), analog outputs (AO), and pulse outputs (PO)

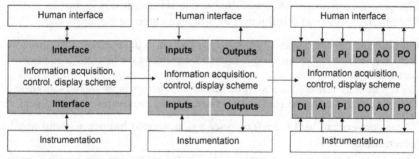

DI: Digital input, AI: Analog input, PI: Pulse input
DO: Digital output, AO: Analog output, PO: Pulse output

Figure 4.2 Interface mechanism.

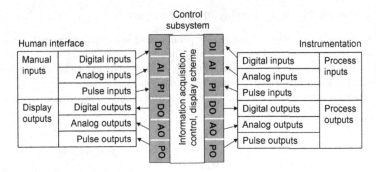

Figure 4.3 Information acquisition and control system.

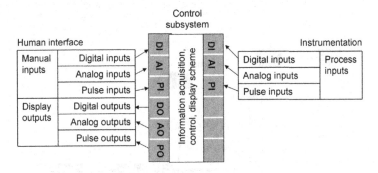

Figure 4.4 Information acquisition system.

Within this structure, there are six input/output-based interfaces for each human interface and instrumentation subsystem, as shown in Figure 4.2. This explains the structure of interfacing at different levels. The requirements for the type and number of inputs and outputs are application specific (flexible and modular).

The general-purpose control subsystem or information acquisition and control system is illustrated in Figure 4.3.

A variant of this general-purpose control subsystem is the information acquisition system and display system, in which there is provision only for information acquisition and display (no provision for process control), as shown in Figure 4.4.

The next question is how we interface the control subsystem with the instrumentation and human interface subsystems.

4.3.2   Instrumentation Subsystem

To illustrate the interfacing of the control subsystem with the instrumentation subsystem, let's get back to the example of the water heating process.

The scenario in Figure 4.5 employs the following:

- *Analog input instrumentation devices*: temperature and level measurements—for acquisition of temperature and level values as analog inputs.
- *Analog output instrumentation devices*: variable control valve and variable voltage drive—for control of water flow and power supply as analog outputs.
- *Integrating instrumentation devices*: water and energy meters—for acquisition of water and energy consumption as pulse inputs.

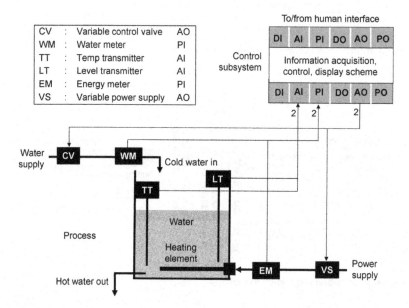

Figure 4.5 Interfacing of process signals—case 1.

Figure 4.5 also indicates the number of input/output (I/O) signals between the instrumentation and control subsystems.

The scenario in Figure 4.6 employs the following:

- *Switching instrumentation devices*: temperature and flow switches—for the acquisition of the status of temperature and level as digital inputs.
- *Digital output instrumentation devices*: on/off control valve and on/off control relay—for the control of water flow and power supply as digital outputs.
- *Integrating instrumentation devices*: water and energy meters—for the acquisition of water and energy consumption as pulse inputs.

Figure 4.6 also indicates the number of I/O signals between the instrumentation and control subsystems.

4.3.3 Human Interface Subsystem

Interfacing the control subsystem with the human interface subsystem is discussed in Chapter 5.

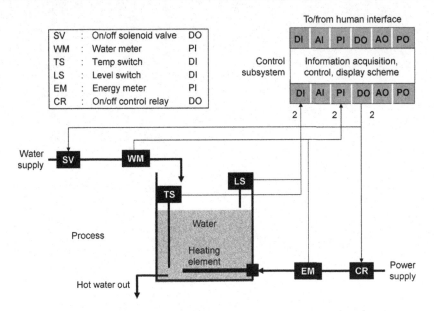

Figure 4.6 Interfacing of process signals—case 2.

4.4 Summary

So far, we have been referring to a block called: information acquisition, control, and display scheme, or **automation strategy**, within the control subsystem. The automation strategy is a mechanism that

- Receives inputs (AI, DI, and PI) via interfaces from instrumentation subsystem for process information acquisition and from human interface subsystem for manual interaction with the process.
- Manipulates the acquired information as per pre-defined criteria.
- Drives outputs (AO, DO, and PO) via interfaces to instrumentation subsystem for process control and to human interface subsystem for process parameter display.

There are many ways of implementing the automation strategy, and these will be further discussed in the forthcoming chapters. The implementation of interfaces within the control subsystem is discussed in Appendix A and Chapter 7.

5 Human Interface Subsystem

5.1 Introduction

In Chapters 3 and 4, we discussed the two basic subsystems, instrumentation and control, and their interfacing. In this chapter, the functions of the human interface subsystem and its interface with the control system are discussed. Further, this chapter deals with the traditional hardwired human interface subsystem, which is popular and cost-effective even in modern automation systems. Chapter 11 discusses the software-based human interface system (operator station).

5.2 Operator Panel

Human interface,[1] also called human–machine interface (HMI) or human–system interface (HSI) or man–machine interface (MMI), is the means by which the operators or users manually interact with the process—they can force an action to control the process or observe the parameters of interest in the process. Another commonly used name for the human interface is **operator panel**.

The basic functions of the human interface system is to directly work with the process to manually control or to assess its state over and above the functions the functions performed by automatically by the control subsystem. The structure of the human interface system facilitates operator control of the functions performed automatically by the control subsystem.

The human interface subsystem is designed with active display and control components mounted suitably on a sheet metal or fiberglass base (panel) and wired up to the control subsystem, as explained in the following sections.

5.2.1 Active Display Elements

These elements are mounted at appropriate locations on the panel for the continuous display of the following:

- Values of analog process parameters, such as of temperature, level, flow (driven by the analog outputs of the control subsystem)
- Status of the digital process parameters, such as of on/off valve, on/off switch (driven by the digital outputs of the control subsystem)

Commonly used active display elements are illustrated in Figure 5.1.

[1] http://www.wikipedia.org.

Overview of Industrial Process Automation. DOI: 10.1016/B978-0-12-415779-8.00005-X

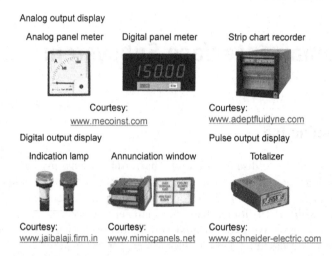

Figure 5.1 Active display elements.

The functions of these active display elements are as follows:

- Both analog panel meters and digital panel meters are functionally the same—they display the current instantaneous values of the continuous parameters. They are driven by the analog outputs of the control subsystem.
- Strip-chart recorders can record the instantaneous values continuously against time and can have more than one pen. This is also driven by the analog outputs of the control subsystem.
- Both the indication lamp and the annunciation window are functionally the same—they display the current status of the discrete parameters. Annunciation windows support the display(s) with a legend with back illumination. The indication lamp and annunciation window are driven by the digital outputs of the control subsystem.
- Totalizers measure and display the counter content by counting the pulses sent out by the pulse outputs from the control subsystem.

In addition to this list, there can be many more active display elements, such as 7-segment displays, hooters, and buzzers.

5.2.2 Active Control Elements

These elements are also mounted at appropriate locations on the panel for the following purposes:

- Manual entry of the set-point values for control of the analog process parameters, such as speed, temperature, flow. This drives the analog inputs of the control subsystem.
- Manual control or change of status of the digital process parameters, such as on/off valve, on/off switch, breaker. This drives the digital inputs of the control subsystem.

Commonly used active control elements are illustrated in Figure 5.2.
The functions of the active control elements are as follows:

- Pressing of the push button switch closes its mechanical contact and allows the voltage to drive the digital input of the control subsystem. The push button switch can be a momentary or latching type. In the momentary type, the digital input to which it is connected

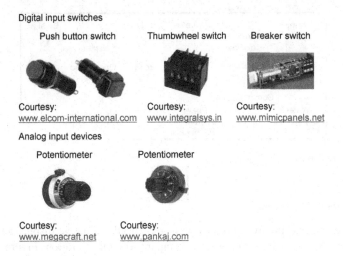

Figure 5.2 Active control elements.

remains active as long as the push button is kept pressed. This is employed for momentary commands. The latching type push button alternates between its two states whenever it is pressed. This is used for setting permanent or regulating commands.

- The thumb wheel switch is used to input the binary-coded-decimal (BCD) digits into the control subsystem. Each BCD digit activates four digital inputs, depending on its combination (0000–1001), by allowing the voltage to activate the digital inputs of the control subsystem as per its binary equivalent (1 is active and 0 is inactive).
- The breaker control switch has three positions (on, neutral, and off). The function (on or off) is similar to the momentary type push button switch. The differences are that the switch generates two momentary commands (one at a time), drives the digital inputs of the control subsystem, and returns to its neutral position after release of the knob.
- The potentiometer is employed to input the set-point/reference values and drives the analog inputs of the control subsystem.

5.2.3 Panel

The panel is a passive element and is just the base of the human interface system, which is made of painted sheet metal, painted fiberglass, or another similar material. The role of the panel is to hold the mounted active components. The active components are arranged on the panel in an order suitable to their operation, wired up, and connected to the control subsystem. The panel and its active components provide easy access to the operator for observing the process parameters of interest (both the states of discrete parameters and the values of continuous parameters) and to effect process control.

5.3 Construction

There are two approaches to constructing the panel—basic and mosaic. These approaches are illustrated in the example of the water heating process. The following sections explain the construction of different types of human interface subsystems.

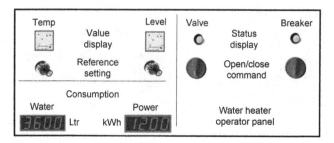

Figure 5.3 Traditional operator panel.

5.3.1 Basic Approach

The basic and cost-effective approach is to mount all the active display and control components on the panel in an order convenient to their operation. Figure 5.3 illustrates this approach for the automation of the water heating process.

This approach allows room for the operator to make a mistake while identifying the active components (display or control), as they do not explicitly link themselves to the process parameters. In other words, the object selection can be ambiguous.

5.3.2 Mimic Approach

The mimic approach overcomes the deficiency in the basic approach. It involves a panel with the process diagram depicted and active control and display components positioned for easy and unambiguous identification. The mimic based operator panel is illustrated in Figure 5.4.

In this approach, there is absolutely no room for the operator to make any mistake while identifying the active components, as they are placed in relation to the process parameters on the panel, and the parameter selection is unambiguous.

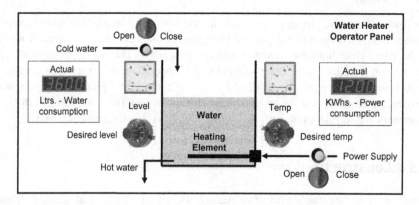

Figure 5.4 Mimic-based operator panel.

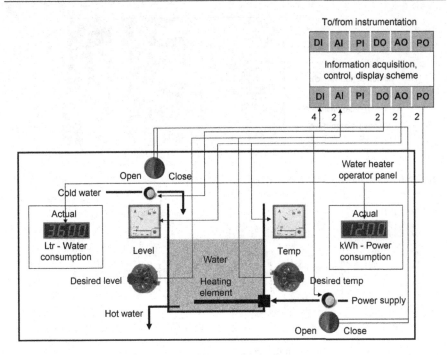

Figure 5.5 Interfacing of mimic panel with control subsystem.

5.4   Interfacing with Control Subsystem

No special interfacing equipment, like instrumentation subsystem, is required between the human interface subsystem and the control subsystem, as they both exchange compatible electronic signals. Figure 5.5 illustrates the interfacing of the mimic panel with the control subsystem. It also indicates the number of input/output (I/O) signals between them.

5.5   Types of Mimic Panels

The hardware-based operator panels can be of two types: sheet metal/fiberglass-based or mosaic tiles-based. There is no difference between sheet metal-based and fiberglass-based except for the panel material. Figure 5.6 illustrates a typical sheet metal-based operator panel in an electrical substation.

The major problem with the sheet metal/fiberglass-based mimic panel is its inflexibility in terms of expanding or modifying the panel at a later stage.

Mosaic tiles-based mimic panels, compared to their sheet metal or fiberglass equivalents, facilitate relatively easy modification or extension of the process diagram for adding more active components. The greatest disadvantage of this panel is its cost. The structure of the mosaic panel is illustrated in Figure 5.7.

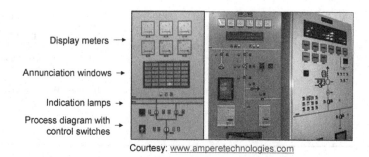

Display meters →
Annunciation windows →
Indication lamps →
Process diagram with → control switches

Courtesy: www.amperetechnologies.com

Figure 5.6 Sheet metal-based mimic panel.

Mosaic tiles   Mosaic mimic/control panel

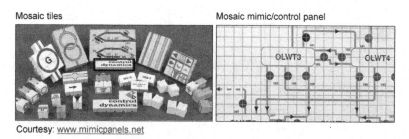

Courtesy: www.mimicpanels.net

Figure 5.7 Mosaic tiles-based mimic panel.

5.6 Summary

In this chapter, we discussed the traditional human interface subsystem and how it interfaces with the control subsystem through standard active display and control components. The major disadvantage of the traditional human interface subsystem (operator panel or mimic-based display/control panel) is that it is inflexibility to expand or modify the panel to meet changed requirements. Even though the mosaic tiles-based panels correct this issue to some extent, they are quite expensive.

6 Automation Strategies

6.1 Introduction

In Chapters 3–5, we discussed the functions of individual subsystems and their associated interfacing. We also mentioned that these subsystems are made to produce the desired result through functions called **information acquisition, control**, and **display scheme**. The chapter on instrumentation subsystems discussed the information acquisition scheme, while the chapter on human interface subsystems discussed the display and manual control scheme. In this chapter, we discuss the function of the control subsystem called the **automation strategy**, which produces the desired results with the support of the other subsystems.

The automation strategy is a predefined and built-in scheme in the control subsystem to guide the automation system to achieve the desired results. It drives the control subsystem to perform information acquisition through instrumentation subsystem and processing as well as process control through the instrumentation subsystem. As per the built-in strategy, the display and manual control functions are achieved through the human interface subsystem. This is also called automation function or automation task. Physical processes, each being different from the other, need specific automation strategies, meeting different requirements to produce different desired results. In other words, each control subsystem is designed to meet a process-specific automation strategy.

Automation strategy operates on input from the process (via the instrumentation subsystem) and from the human interface subsystem (for manual control). It analyzes the information and produces the required command output, as per the predefined criteria, and sends the output to the process (via instrumentation) and to the human interface subsystem for display of process parameters of interest. Figure 6.1 illustrates the functions of information acquisition, control, and display for automation strategy.

The implementation of automation strategies is explained in the subsequent sections using the automation of a water heater as an example.

6.2 Basic Strategies

The basic strategies for the implementation of automation schemes are the following:

- Open loop control
- Closed loop control or feedback control

Overview of Industrial Process Automation. DOI: 10.1016/B978-0-12-415779-8.00006-1

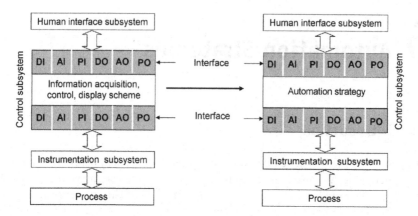

Figure 6.1 Automation strategy.

6.2.1 Open Loop Control

Open loop control strategy supports **pre-known** results (responses) to the control inputs. No assessment of the responses and corrections is possible for any internal and/or external disturbances. This scheme is illustrated in Figure 6.2.

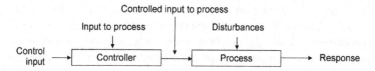

Figure 6.2 Open loop control.

This scheme, though used quite frequently, cannot always assure the desired result. It is simple, economical, and is used in less demanding applications, and it is applied in both discrete and continuous process automation as discussed in forthcoming sections.

6.2.2 Closed Loop Control

Closed loop control, also known as feedback control, eliminates the shortcomings of open loop control. Here, the response or the actual result is continuously compared with the desired result, and the control output to the process is modified and adjusted to reduce the deviation, thus forcing the response to follow the reference. Effects of the disturbances (external and/or internal) are automatically compensated for. This scheme is superior, complex, and expensive. It is used for more demanding applications and is commonly applied in continuous process automation as discussed in forthcoming sections. Figure 6.3 illustrates this scheme.

The following sections discuss the application of these basic strategies in the automation of discrete, continuous, and hybrid processes.

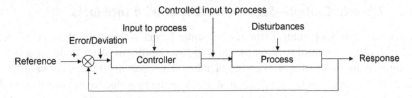

Figure 6.3 Closed loop control.

6.3   Discrete Control

Discrete control is employed for processes involving only discrete inputs and discrete outputs and their associated instrumentation devices. The discrete control can be further classified into open loop control and sequential control with interlocks.

6.3.1   Discrete Control—Open Loop

On/off commands are issued to produce the desired results for open loop discrete control. This scheme does not compensate for disturbances. Figure 6.4 illustrates the open loop discrete control scheme.

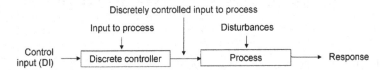

Figure 6.4 Discrete control—open loop.

Figure 6.5 illustrates the application of open loop discrete control for the control of temperature and level in a water heater. This is nothing more than an on/off control of the valve and the breaker.

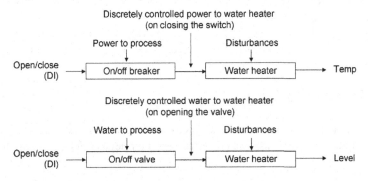

Figure 6.5 Open loop control of water heater (discrete).

6.3.2 Discrete Control—Sequential Control with Interlocks

In a discrete open loop control strategy for water heater automation, there is one serious drawback: the control strategy assumes the availability of water in the tank before switching on the power for the heating element. This may not always be the case for several reasons, such as no or insufficient water in the tank due to clogged water inlet pipe. If this is not taken care of, there is a possibility for a heating element to get damaged. Therefore, it is necessary to allow sufficient time for the water to fill up before switching the power on. This strategy becomes safer if the power gets switched on only after ascertaining the water level in the tank.

Sequential control with interlocks addresses the drawbacks in a simple open loop control. Like a discrete open loop control, the instrumentation devices for both the data acquisition and control are discrete. Sequential control with interlocks ensures, in each step, the desired intermediate conditions or **interlocks** are satisfied before executing the next step. Figure 6.6 illustrates sequential control with interlocks.

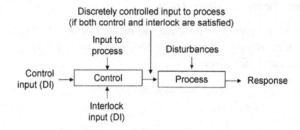

Figure 6.6 Discrete control—sequential control with interlocks.

Figure 6.7 illustrates the application of sequential control with interlocks for a water heater. It makes sure there is sufficient water in the tank before allowing the heating element to be turned on.

As seen in Figure 6.7, the breaker does not get closed—even if a command is given (either manually or automatically)—unless the water level is full. The control moves in steps, or sequentially, but only after satisfying certain conditions at every step until the desired result is reached. Here also, the instrumentation devices for both data acquisition and control are discrete. Interlocks are predominantly used for meeting safety requirements.

Sequential control with interlocks is widely used for startup/shutdown of complex plants. Some common examples of its use can be seen in the operation of passenger lifts and traffic signals.

6.4   Continuous Control

Continuous control is for continuous processes, and it employs analog inputs and analog outputs with their associated instrumentation devices. This strategy has two variants:

- Open loop control
- Closed loop or analog loop control

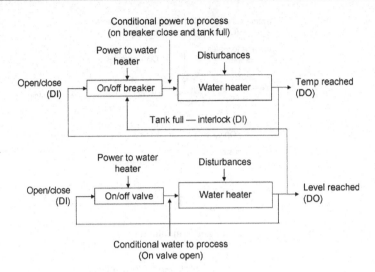

Figure 6.7 Sequential control with interlocks of water heater.

6.4.1 Continuous Control—Open Loop

Figure 6.8 illustrates the simple open loop continuous control. In this case, the response is proportional to the input. To achieve the desired result, the input to the process must simply be varied. Since it is open loop control, it does not compensate for the disturbances.

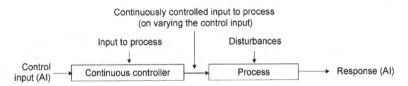

Figure 6.8 Continuous control—open loop.

Figure 6.9 illustrates the application of simple open loop continuous control for the level and temperature of a water heater. Here, the variable voltage source increases or decreases the power flow to the heating element proportional to the input. Similarly, the variable control valve increases or decreases the water flow to the tank proportional to the input.

6.4.2 Continuous Control—Closed Loop

For closed loop control, or **analog loop control**, the need is to continuously track the process output, compare this with the reference or the desired output, and vary the control input proportionally to minimize the deviation or the error (output to follow the reference).

Here, both the data acquisition and control and their associated instrumentation devices are continuous. Figure 6.10 illustrates the scheme of closed loop control.

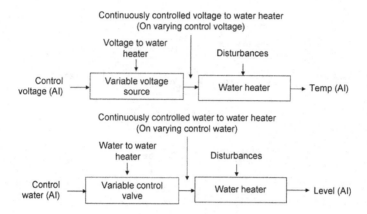

Figure 6.9 Open loop control of water heater (continuous).

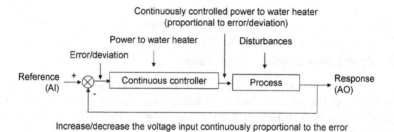

Figure 6.10 Continuous control—closed loop.

Figure 6.11 illustrates the application of closed loop control for the temperature and level of a water heater. With the water heater as the process and the variable voltage source and variable control valve as controllers, the actual temperature and level are continuously measured and compared with their desired values to generate the deviations. These deviations proportionally increase or decrease the power input to the heating element and the water input to the tank forcing the process to follow the reference values.

6.5 Hybrid Control

Hybrid processes are a combination of both discrete and continuous processes. The control schemes are discussed in the following sections.

6.5.1 Hybrid Control—Two-Step

Two-step control is a crude approach to continuous closed loop control, and it employs continuous inputs for information acquisition and produces discrete outputs for control execution. This is very similar to the functioning of a switching

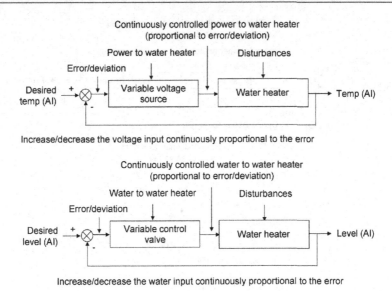

Figure 6.11 Closed loop control of water heater (continuous).

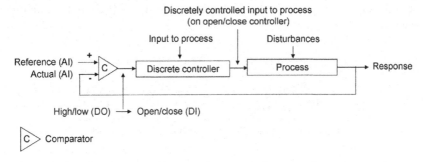

Figure 6.12 Two-step control.

instrumentation device in the instrumentation subsystem. The only difference is that, in the switch, the reference value is set locally in the instrumentation device itself, while in the two-step control, the reference value is set in the human interface subsystem. Just as with the switch, the control is exerted discretely only if the process parameter deviates from the reference input. This is also called **on/off** or **bang-bang** control. The information acquisition and its associated instrumentation devices are continuous (analog), while the control and its associated instrumentation devices are discrete (digital). Figure 6.12 illustrates the two-step control scheme.

Figure 6.13 illustrates the application of two-step control for temperature and level in a water heater.

Two-step control, being a crude approach to continuous control, has a serious downside. The control command operates on the final control elements (valve and breaker) even for very minor deviations, forcing them to hunt or oscillate between

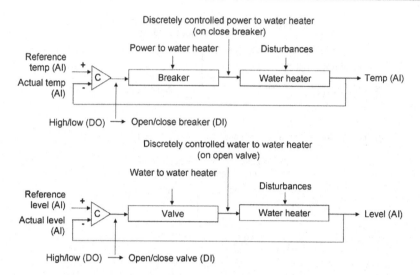

Figure 6.13 Two-step control of water heater (hybrid).

their two positions around the reference value. As illustrated in Figure 6.14, this causes a lot of wear and tear on the final control elements, especially the electromechanical ones.

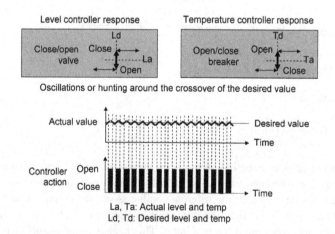

Figure 6.14 Two-step control—performance.

6.5.2   Hybrid Control—Two-Step with Dead-Band

To a major extent, the hunting or oscillation problem can be reduced by introducing a **dead-band** in the control scheme for taking action only when the process value goes outside the preset dead-band. In other words, in two-step control with a dead-band, the process output is always forced to stay within the dead-band. The lower the dead-band, the oscillations will be higher and the control will be finer. The higher the dead-band, the oscillations will be lower and the control will be coarser. **No correction is done within the dead-band**.

Figure 6.15 illustrates the two-step control scheme with dead-band.

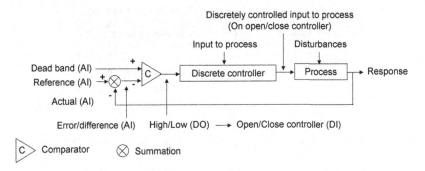

Figure 6.15 Two-step control with dead-band.

Figure 6.16 illustrates the application of two-step control with a dead-band for temperature and level control in a water heater.

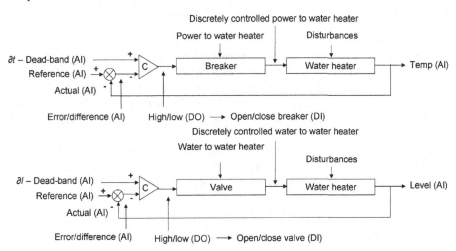

Figure 6.16 Two-step control with dead-band of water heater.

Here, ∂l is the dead-band for the level, and the level controller closes the valve only when the actual level goes above ***desired level*** $+\partial l$. It opens the valve when the actual level goes below ***desired level*** $-\partial l$.

Similarly, with ∂t as the dead-band for temperature, the temperature controller opens the switch only when the actual temperature goes above desired ***temperature*** $+\partial t$. It closes the switch when the actual temperature goes below desired ***temperature*** $-\partial t$.

Figure 6.17 illustrates the performance of the two-step control with a dead-band. As seen here, the lower the dead-band, the finer the control but with less stability. Conversely, the higher the dead-band, the coarser the control but with higher stability.

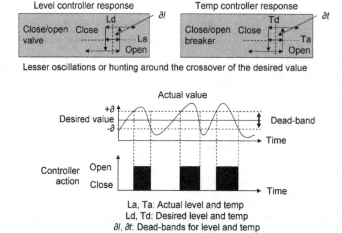

Figure 6.17 Two-step control with dead-band—performance.

Typical examples of two-step control with a dead-band are room air conditioners and refrigerators where the preset band is factory set.

Theoretically, two-step control without any dead-band or with a very small dead-band can produce almost continuous control, but this scheme is not feasible due to oscillatory or unstable response leading to frequent switching operations in the final control elements.

Therefore, two-step control with a dead-band does not perform a true continuous control even with the decrease of dead-band. It produces undesired effects on the final control elements because of inherent hunting or oscillation.

6.6 Summary

In this chapter, we described various control strategies of the automation system. The advantages and disadvantages of strategies for the automation of discrete, continuous, and hybrid processes were discussed using the example of a water heating process. This chapter furthers the discussion on the structure and design of the control subsystem. This is explained more in Appendix A and in Chapter 7.

7 Programmable Control Subsystem

7.1 Introduction

In their early days, automation systems used pneumatic, mechanical, and hydraulic components for both measurement and control. These were all slow, bulky, sluggish, less reliable, and they required more maintenance, space, power. Automation technology gradually moved toward electrical components and is now based on electronics.

As explained in Appendix A, modern control systems started out as hardwired systems, utilizing the technology that was available at the time. They began with relay technology, followed by solid state technology (The reader is recommended to go through Appendix A before proceeding further). Later, automation technology moved into **processor-based** (microprocessors or computers) or information technology. Advances in electronics, information, communication, and networking technologies have played a vital role in making the entire control subsystem programmable, more compact, power-efficient, reliable, flexible, communicable, and self-supervisable. All the drawbacks, as discussed in Appendix A, associated with the hardwired control subsystem are overcome with a **programmable control subsystem** or **soft-wired control subsystem**, which also brings many more advantages.

The example of the water heating process, with its instrumentation and human interface subsystems, is employed throughout this chapter to discuss the implementation of various types of control subsystems with processor-based technology.

Figure 7.1 illustrates the migration of a general-purpose **hardwired** control subsystem into a general-purpose processor-based, or **soft-wired**, control subsystem.

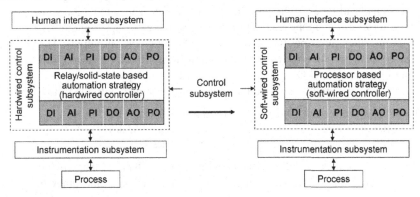

Figure 7.1 Transition of hardwired controller to soft-wired controller.

Overview of Industrial Process Automation. DOI: 10.1016/B978-0-12-415779-8.00007-3

This control subsystem has all the advantages of a solid state control subsystem, as the processor-based systems are also solid state. In addition, they have the advantages of modularity, extendibility, and flexibility for strategy modification or extension (soft-wired automation strategy). Since they are soft-wired, the connections are established through software. However, the soft-wired control subsystem requires programming skills (explained in Chapter 10) to realize the automation strategy.

In the subsequent sections and chapters, the programmable control subsystem will be referred to as the **controller**. The specific industrial names of different types of controllers, based on their application domains, are discussed in Chapter 13.

The information structure in the controller is illustrated in Figure 7.2. During information acquisition, electronic inputs to the control subsystem are converted by the controller into computer data or simply **data** for further processing within the controller. Similarly, during control execution, processed information in the form of data is converted into electronic output by the controller.

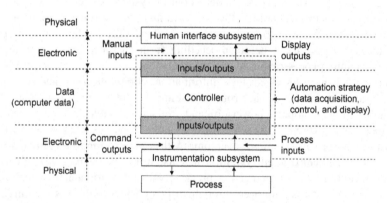

Figure 7.2 Information structure in controller.

Figure 7.3 illustrates the structure of the general-purpose controller with a provision to support all types of process inputs and outputs. The automation strategy here is in the form of a memory resident program operating on inputs and other memory resident information, as per the strategy to produce the control outputs. The controller can be designed for any type of automation strategy with the appropriate combination of input/output (I/O) modules and automation strategy.

Like a typical computer system, the controller is designed with the following:

- Power supply subsystem consisting of a power supply module
- Processor subsystem consisting of processor and memory modules
- I/O subsystem consisting of various types of I/O modules

All these modules are physically integrated over a **bus**, which provides a path for communication and data exchange between the processor and other functional modules.

As each I/O channel within an I/O module is logically independent, instrumentation subsystems and human interface subsystems can share a single I/O module with multiple channels, if required. The memory resident strategy takes care of the distinction. This is illustrated in Figure 7.3.

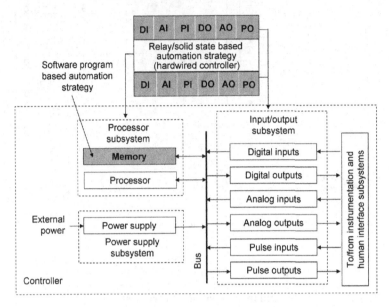

Figure 7.3 Controller structure.

The following sections describe the implementation of various control strategies (discrete as well as continuous control types) using the example of a water heating process. The implementation of the automation strategy using relay and solid state systems (see Appendix A) is re-implemented here with the controller.

7.2   Discrete Control

As already discussed, the discrete process needs only discrete inputs and outputs to support either open loop control or sequential control with interlocks. In both strategies, hardware configuration of the controller remains identical. However, software for the automation strategy is different in each case. Hence, in the following section, only the sequential control with interlocks is discussed.

7.2.1   Sequential Control with Interlocks

Figure 7.4 illustrates the water heating process with an instrumentation subsystem with discrete devices and a human interface subsystem with discrete panel components.

The functions and interconnections of the instrumentation devices and the human interface components with the controller are explained in Table 7.1.

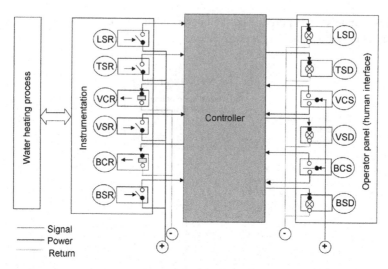

Figure 7.4 Water heater with automation system (discrete).

Table 7.1 List of Instrumentation Devices

Between controller and instrumentation (process)
Discrete instrumentation devices

1	LSR	Level supervision relay (preset value reached/not reached)
2	TSR	Temp supervision relay (preset value reached/not reached)
3	VCR	Valve control relay (command to open/close valve)
4	BCR	Breaker control relay (command to close/open breaker)
5	VSR	Valve supervision relay (opened/closed)
6	BSR	Breaker supervision relay (closed/opened)

Between controller and human interface (operator panel)
Discrete panel components

1	LSD	Level status indication lamp (on/off for reached/not reached)
2	TSD	Temp status indication lamp (on/off for reached/not reached)
3	VCS	Valve control switch (momentary command to open/close)
4	BCS	Breaker control switch (momentary command to close/open)
5	VSD	Valve status indication lamp (on/off for opened/closed)
6	BSD	Breaker status indication lamp (on/off for opened/closed)

Figure 7.5 illustrates the adaptation of the general-purpose controller to support discrete process automation with only digital inputs and outputs with the program for sequential control with interlocks for the water heating process.

Table 7.2 illustrates the process I/O signal allocation to inputs and outputs of the controller.

Along with the allocation of process signals to different I/O channels, the strategy also requires some memory locations for remembering the momentary conditions

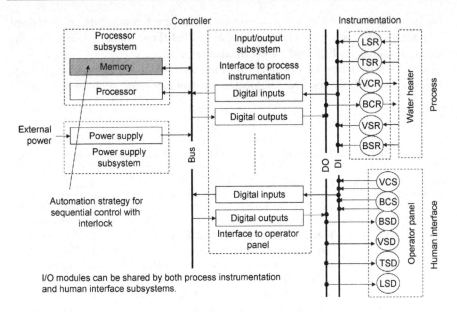

Figure 7.5 Discrete process automation—controller structure.

Table 7.2 Allocation of I/O Channels to Process Parameters

From/To	I/O Channel	Instrumentation Device		Process Signal

Digital input module (8 channels: DI0–DI7)

From/To	I/O Channel	Instrumentation	Device	Process Signal
From process	DI0	LSR	Level supervision relay	Level status
	DI1	TSR	Temp supervision relay	Temp status
	DI2	VSR	Valve supervision relay	Valve status
From operator	DI3	BSR	Breaker supervision relay	Breaker status
panel	DI4	VCS	Valve control switch—open	Open valve command
	DI5	VCS	Valve control switch—close	Close valve command
	DI6	BCS	Breaker control switch—open	Open breaker command
	DI7	BCS	Breaker control switch—close	Close breaker command

Digital output module (8 channels: DO0–DO7)

From/To	I/O Channel	Instrumentation	Device	Process Signal
To operator	DO0	LSD	Level status indication	Level status
panel	DO1	TSD	Temp status indication	Temp status
	DO2	VSD	Valve status indication	Valve status
	DO3	BSD	Breaker status indication	Breaker status
To process	DO4	VCR	Valve control relay	Open/close valve
	DO5	BCR	Breaker control relay	Close/open breaker
–	DO6–DO7	Not used		

(similar to latching of momentary commands into level commands, as explained in Appendix A). These memory locations are called **flags**. Table 7.3 describes the memory locations for flags and for intermediate results.

Figure 7.6 illustrates the flow chart for the execution of the automation strategy. The program memorizes the momentary commands (for momentary control

Table 7.3 Allocation of Memory Locations for Special Requirements

Name	Location	Function	State
Flags	FVO	To remember issue of momentary valve open command (flag valve open)	0—No command 1—Command
	FVC	To remember issue of momentary valve close command (flag valve closed)	0—No command 1—Command
	FBO	To remember issue of momentary breaker open command (flag breaker open)	1—Command 1—Command
	FBC	To remember issue of momentary breaker close command (flag breaker closed)	0—No command 1—Command
	XXX	To store the intermediate results temporarily	

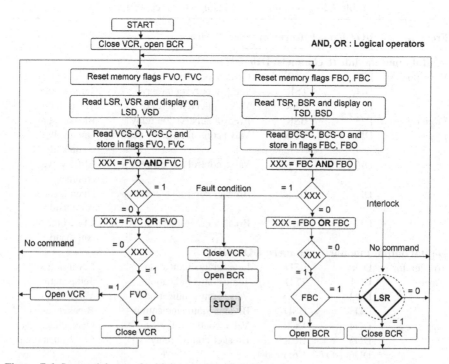

Figure 7.6 Sequential control with interlocks—flow chart.

commands), interlock management, etc. The following steps must be applied before the flow chart becomes executable by the controller:

- Code the flow chart into a program using an automation application programming language.
- Convert the automation program into a machine executable code.
- Transfer the machine executable code into the memory of the controller.

With the completion of these steps, the controller is ready to execute the automation strategy, and it **behaves** like a hardwired control subsystem until the strategy is reprogrammed and reloaded.

The formal application programming languages used for writing industrial automation strategies and related steps are covered in Chapter 10.

7.3   Continuous Control

As discussed earlier, the continuous process needs only continuous inputs and outputs (continuous control) to support both open loop and closed loop control. Hardware configuration of the controller is the same for these two cases except for the number of I/O channels and the automation strategy. In view of this, only closed loop control is discussed here.

7.3.1   Closed Loop Control

Figure 7.7 illustrates the water heating process with an instrumentation subsystem with analog devices and a human interface subsystem with analog panel components.

The functions and interconnections of the instrumentation devices and human interface panel components with the controller are given in Table 7.4.

Figure 7.8 illustrates the controller structure and its interfaces to both instrumentation and human interface subsystems.

Table 7.5 illustrates the allocation of I/O channels of the controller to the process signals. Along with the allocation of process signals to different I/O channels, the program also requires some memory locations for storing the intermediate results. Table 7.6 describes the memory locations for intermediate results.

Figure 7.9 illustrates the flow chart for the execution of the automation strategy.

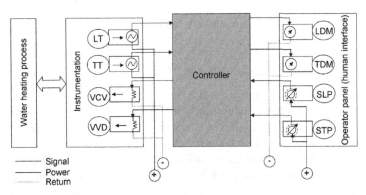

Figure 7.7 Water heater with automation system (continuous).

Table 7.4 List of Instrumentation Devices

Between controller and instrumentation (process)
Analog instrumentation devices

1	LT	Water level transmitter
2	TT	Water temp transmitter
3	VCV	Variable control valve
4	VVD	Variable voltage drive

Between controller and human interface (operator panel)
Analog panel components

1	LDM	Water level indication meter
2	TDM	Water temp indication meter
3	SLP	Desired water level setting potentiometer
4	STP	Desired water temp setting potentiometer

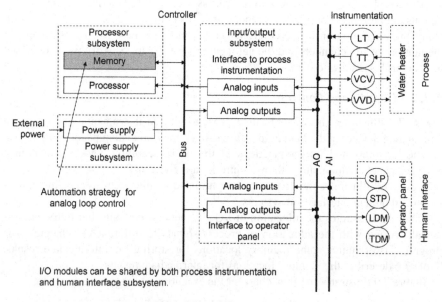

Figure 7.8 Continuous process automation—controller structure.

7.3.2 Multi-Input/Multi-Output Control

The automation of the continuous water heating process discussed so far has two closed loops—one for level control and the other for temperature control. In this example, each loop is independent and does not require any interaction with the other loop. Each loop is an example of a single-input/single-output (SISO) controller. However, in practice, continuous processes have multiple loops operating parallel to each other, and this requires interaction among them. This is a case of a

Table 7.5 Allocation of Process Signals to I/O Channels

From/To	I/O Channel	Instrumentation Device		Process Signal
Analog input module (8 channels: AI0–AI7)				
From process	AI0	LT	Level transmitter	Water level
	AI1	TT	Temp transmitter	Water temp
From operator panel	AI2	SLP	Set level potentiometer	Desired water level
	AI3	STP	Set temp potentiometer	Desired water temp
–	AI4–AI7	Not used		
Analog output module (4 channels: AO0–AO3)				
To process	AO0	VCV	Variable control valve	Level status
	AO1	VVD	Variable voltage drive	Temp status
To operator panel	AO2	LDM	Level indication meter	Valve status
	AO3	TDM	Temp indication meter	Breaker status

Table 7.6 Allocation of Memory Locations for Special Requirements

Name	Location	Function	State
	XXX	To store the intermediate results temporarily	

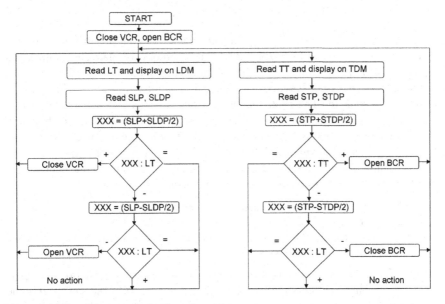

Figure 7.9 Closed loop control—flow chart.

multi-input/multi-output (MIMO) process, and the controller is a multi-input/multi-output controller. Figure 7.10 illustrates the schematics of both SISO and MIMO loops. The MIMO controller provides for control outputs as functions of one or more inputs.

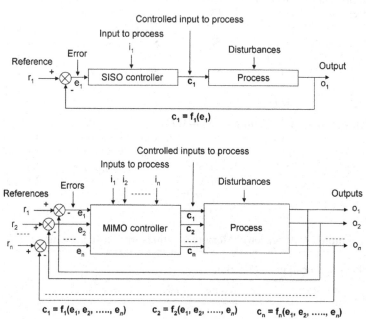

Figure 7.10 SISO and MIMO control subsystem.

Unlike discrete process automation, implementation of control subsystems for the MIMO continuous process using the hardwired technology, is relatively complex and may not even be possible if the number of interacting loops is large. However, the controller, being processor-based, can easily be programmed to work as a MIMO control subsystem since the data related to all the loops is available in the common memory for use by any loop. Furthermore, the processor in the controller, which is very fast, can handle many loops simultaneously while still meeting the time requirements (the timing and duration within which each loop is allowed to complete its task).

To illustrate this, let's discuss the design of a 2 input–2 output controller for the water heating process with the following requirements:

- *Loop 1*: Temperature control to maintain the temperature at the reference value.
- *Loop 2*: Level control to maintain the level at the reference value.
- *Interaction between loop 1 and loop 2*: Temperature control works only when the water level is above 25% of the reference level.

The setup of the continuous water heating process with instrumentation and interface or a controller configuration (Figures 7.7 and 7.8) remains the same. Only the strategy implementation changes, as illustrated in the flow chart in Figure 7.11. Here, the temperature control loop (loop 2) is supervised by the level control loop (loop 1).

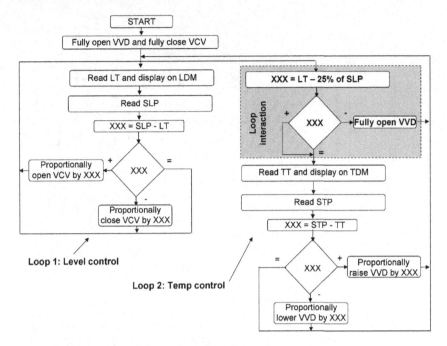

Figure 7.11 2 input–2 output controller—flow chart.

7.4 Hybrid Control

As mentioned in Chapter 6, hybrid control has the following two versions:

- Two-step control
- Two-step control with dead-band

Two-step control is a combination of both discrete and continuous control philosophies, and it is a good and cost-effective approximation of the true continuous control implemented by a closed loop control.

As the controller hardware configuration is the same for both two-step and two-step with dead-band control, only the latter is discussed here.

7.4.1 Two-Step Control with Dead-Band

Figure 7.12 illustrates the water heating process with its analog and digital instrumentation devices and analog human interface panel components.

The list of the analog and digital instrumentation devices and the analog human interface panel components are given in Table 7.7.

Figure 7.13 illustrates the controller configuration for the implementation of two-step control with dead-band for the water heating process.

Table 7.8 illustrates the allocation of process inputs and outputs of the controller while Table 7.9 illustrates the allocation of memory location for intermediate results.

Figure 7.14 illustrates the flow chart for the execution of the automation strategy.

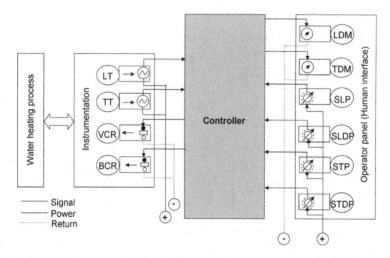

Figure 7.12 Water heater with automation system (hybrid).

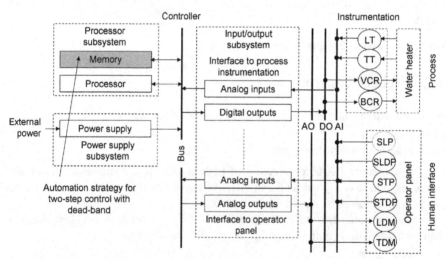

I/O modules can be shared by both process instrumentation and human interface subsystems.

Figure 7.13 Hybrid control—controller structure.

7.5   Controller with Additional Features

Communicability and **self-supervision** are unique features of the processor-based systems. These features are also implemented in the controller and are discussed in the following sections.

Table 7.7 List of Instrumentation Devices

Between controller and instrumentation (process)
Analog instrumentation devices

1	LT	Water level transmitter
2	TT	Water temp transmitter

Discrete instrumentation devices

1	VCR	Valve control relay
2	BCR	Breaker control relay

Between controller and human interface (operator panel)
Analog panel components

1	LDM	Water level indication meter
2	TDM	Water temp indication meter
3	SLP	Desired water level setting potentiometer
4	STP	Desired water temp setting potentiometer

Table 7.8 Allocation of Process Signals to I/O Channels

From/To	I/O Channel	Instrumentation Device		Process Signal
Analog input module (8 channels: AI0–AI7)				
From	AI0	LT	Water level transmitter	Water level
process	AI1	TT	Water temp transmitter	Water temp
From operator	AI2	SLP	Set level potentiometer	Desired water level
panel	AI3	SLDP	Set level dead-band potentiometer	Desired water temp
	AI4	STP	Set temp potentiometer	Desired level dead-band
	AI5	STDP	Set temp dead-band potentiometer	Desired temp dead-band
Not used	AI6–AI7	–		
Analog output module (4 channels: AO0–AO3)				
To operator	AO0	LSD	Water level indication meter	Level status
panel	AO1	TSD	Water temp indication meter	Temp status
Not used	AO2–A03	–		
Digital output module (8 channels: DO0–DO7)				
To process	DO0	VCR	Valve control relay	Open/close valve
	DO1	BCR	Breaker control relay	Close/open breaker
Not used	DO2–DO7	–		

Table 7.9 Allocation of Memory Locations for Special Requirements

Name	Location	Function	State
	XXX	To store the intermediate results temporarily	

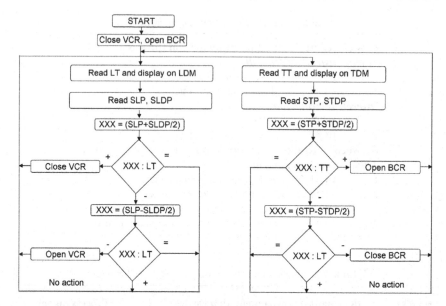

Figure 7.14 Two-step control with dead-band—flow chart.

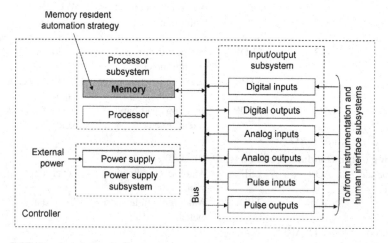

Figure 7.15 Stand-alone controller.

7.5.1 Communicability

So far, we discussed controllers that can only send and receive **analog electronic signals** to and from the instrumentation and human interface subsystems. Hence, they are called **stand-alone controllers**. See the illustrated example in Figure 7.15.

As illustrated in Figure 7.16, controllers are required to be communicable, so they must have a communication interface module for communication with compatible external systems, devices, and equipment, over communication media. This is called

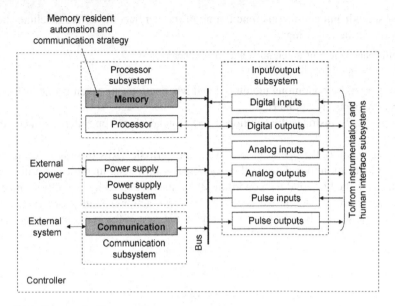

Figure 7.16 Communicable controller.

a **communicable controller** and has both the automation strategy and communication strategy in its memory. Communicability means that the system can communicate with the external world for **data exchange**.

A good example of the utility of the communicable controller can be seen in a remotely placed coffee machine built with stand-alone and communicable controllers.

The starting conditions and sequence of operations involved in making the coffee are as follows:

- The coffee machine is placed in a remote location and is supervised by the operator located in his workplace.
- Bins are full with roasted coffee beans, milk, and water.
- The coffee machine is healthy and properly functioning (fault-free).
- The user places a cup below the tap, presses the start button, and waits for the cup to get filled with coffee.

The automated coffee machine executes the following steps sequentially after the start button is pressed:

1	Start	Grinds the required quantity of roasted coffee beans to produce the coffee powder
2		Heats the water to the required temperature
3		Passes the water through the freshly ground coffee powder to produce the decoction
4		Collects the required quantity of the decoction
5		Adds the required quantity of milk to the decoction
6		Heats the mixture to the required temperature
7	Stop	Fills the cup

For smooth and continuous functioning of this service, the coffee machine should always have the following:

• Sufficient quantity of water, milk, and roasted beans
• Proper working condition with no faults

How are these conditions monitored and what actions are taken if one or more of the requirements are not met?

Figure 7.17 illustrates the solution with the stand-alone controller-operated coffee maker.

In this solution, the operator needs to make periodic visits from his workplace to the remote location to make sure the machine has sufficient quantities of milk, water, and beans in the bins, and the condition of the machine (healthy or faulty). The operator takes corrective actions, if necessary (reactive actions). With more visits by the operator, the service becomes better, but the efficiency of the operator becomes poor.

Remote place

Figure 7.17 Stand-alone controller coffee maker.

Figure 7.18 illustrates the solution with a communicable controller-operated coffee maker.

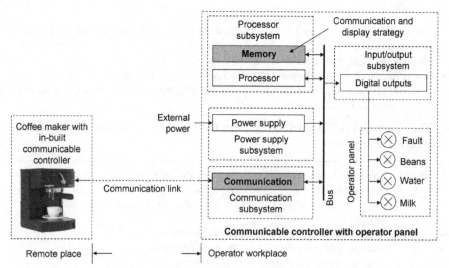

Figure 7.18 Communicable controller coffee maker.

As the communicable controller-based coffee maker can communicate remotely with the operator's workplace, the operator gets continuous online information on the status of the raw material availability in the bins and the machine status (healthy or faulty) at his work place itself. The operator takes proactive or advance actions, if required.

The communicable controller solution is expensive and technically more complicated, but it provides a better and more efficient service. Further, this solution can be extended to support multiple coffee makers distributed in remote places but connected to a single operator panel in the operator's workplace over the communication line.

7.5.2 Self-Supervision or Watchdog

Self-supervision means the ability of the processor-based system to continuously diagnose its own health and announce any abnormality. To facilitate this, the system has a built-in watchdog function. Watchdog functionality (a combination of hardware and software) monitors and announces both **fatal** and **nonfatal failures**. Nonfatal failures do not completely knock off the controller while fatal failures totally knock off the controller. Figure 7.19 illustrates a controller with a watchdog facility.

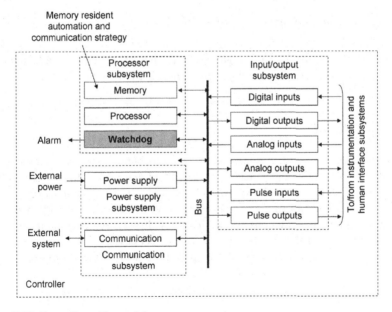

Figure 7.19 Controller with watchdog.

7.6 Upward Compatibility

Normally, the process equipment, instrumentation subsystem, and human interface subsystem do not become obsolete quickly and last much longer than the control subsystems. There is often a need to change the control subsystem for a variety of reasons: to increase its functionality, to extend its I/O capacity, because of the unavailability of spares/service due to obsolescence, etc. The old technology might become obsolete or unable to meet the additional functionality, I/O increase, performance, etc. To provide for investment protection, the new generation control subsystems are designed to replace the older generation systems as long as the new generation system's terminal configurations, functionality, physical parameters, etc., are made compatible to the replaced ones (upward compatibility). This is illustrated in Figure 7.20.

Terminals for input/output connections to human interface subsystem

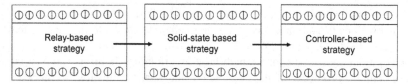

Terminals for input/output connections to instrumentation subsystem

ⓘⓘⓘⓘ Screw terminals

Figure 7.20 Upward compatibility of control subsystems.

In this case, controllers score high marks, as they are compact, modular in I/O, programmable, and easily adaptable to any changed requirements.

7.7 Summary

A general-purpose controller is modular, programmable (soft-wired), communicable with all types of I/O modules, and has customized automation functions (software for data acquisition, control, display, and communication). The functions include open loop control and all variants of closed loop control with multiple I/O, with multiple loops, and with interactions among the loops, display, and communication. This general-purpose controller can be customized to meet any specific automation applications. This chapter also discussed the special features of the controller, namely, communicability and self-supervision.

8 Hardware Structure of Controller

8.1 Introduction

In Chapter 7, we discussed the philosophy of the controller with specific reference to the implementation of various automation, display, and communication strategies. This chapter details the hardware construction of the controller in general and its functional modules in particular.

A general-purpose controller, as illustrated in Figure 8.1, is computer-like equipment and functions like a computer with all the necessary functional modules. A general-purpose computer employs devices like keyboards and displays as I/O devices, while the controller employs process I/O modules as I/O devices.

Figure 8.2 illustrates the typical hardware structure of the controller. The structure and the functions of the various modules are described in the following sections. In practice, many different arrangements are available depending on the manufacturer. The one discussed here is a simple, early implementation to best explain the concepts involved.

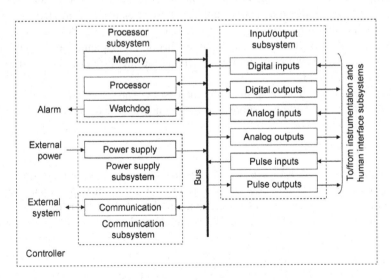

Figure 8.1 Controller—logical structure.

Overview of Industrial Process Automation. DOI: 10.1016/B978-0-12-415779-8.00008-5

Front view

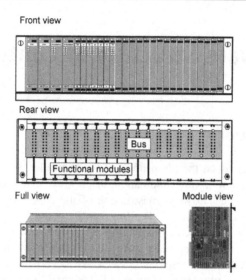

Rear view

Full view   Module view

Figure 8.2 Controller—physical structure.

8.2 Major Modules of Controller

8.2.1 Rack

A rack is a mechanical structure that holds the functional modules in their places to facilitate their physical and electrical connections to the bus. The general arrangement of the rack is illustrated in Figure 8.3. A rack can hold a limited number of functional modules. Hence, to accommodate more modules, additional racks are needed. Racks are of two types: main and extension or supplementary. Even though they look similar, they are structurally different, and the difference between them is covered in later sections.

Some slots in the main rack (normally the first few) are reserved for specific functional modules, such as power supply, processor, memory, and communication.

8.2.2 Bus

The bus, as illustrated in Figure 8.4, is a passive electronic assembly used to supply power to the functional modules and also to provide a communication path between the processor module and the other functional modules for data transfer.

As illustrated in Figure 8.5, a typical bus in its simplest form has lines or tracks to carry the following:

- Power to all functional modules (power lines)
- Addresses of memory and functional modules (address lines)
- Data in both directions (data lines)
- Read and write control (control lines)
- Interrupt and, clock (special lines)

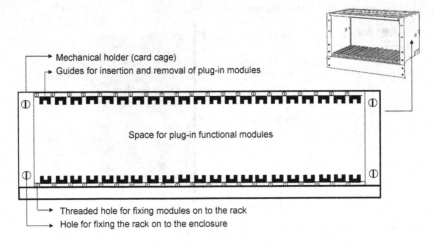

Figure 8.3 Rack.

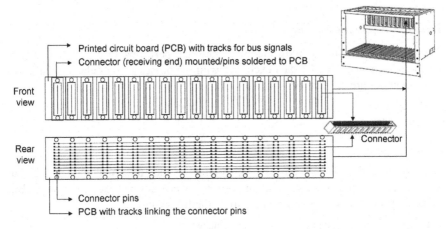

Figure 8.4 Bus—physical structure.

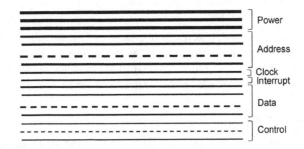

Figure 8.5 Bus—logical structure.

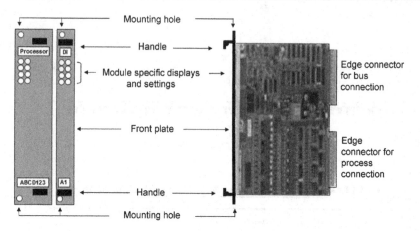

Figure 8.6 Functional module.

The number of lines or tracks on the bus in each category depends on the architecture of the processor (microprocessor/computer) employed in the controller.

8.2.3   Functional Modules

The functional module, as illustrated in Figure 8.6, performs specific functions and has the bus connectivity for data exchange with the processor module. Typical functional modules are as follows:

- Memory
- Input/output
- Communication
- Watchdog

As illustrated in Figure 8.6, the top portion of the functional module has bus interface electronics, while the bottom portion has process interface electronics. Bus interface electronics enable the module to do the following:

- Communicate with the processor for data exchange
- Temporarily store the information related to the module before it is either passed on to the process or passed on to the processor module
- Control and supervise the operation of the module
- Diagnose the fault in the module

Process interface electronics facilitate interfacing of the module with the process signals (via instrumentation devices) either for converting the incoming electronics signals to data or for converting the outgoing data to electronic signals.

More discussion on the bus and process interfacing can be found in Appendix B.

8.2.4   System Cable

As illustrated in Figure 8.7, the system cable is a multi-core signal cable with a connector on one end for connecting to the functional module and a terminal block at

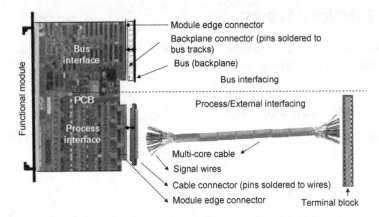

Figure 8.7 System cable.

the other end for facilitating the wiring to the instrumentation devices and/or human interface components. The number of cores depends on the number of connections that are present on the functional modules. Generally, the system cables are prefabricated to terminate the incoming/outgoing signals to and from the functional modules within the enclosure of the controller.

8.3 Data Exchange on Bus

To explain these basic concepts, the processor module can be seen as the master of the bus, while all the other functional modules, such as I/O, communication, and watchdog, are the slaves on the bus. The data exchange on the bus is totally controlled by the processor module. The data exchange sequences are as given below.

From the slave module to the processor module:

- Processor module places the address of the slave module on the bus.
- Processor module places the read control signal on the bus.
- Addressed slave module places the data on the bus.
- Processor module accepts the data from the bus.

From the processor module to the slave module:

- Processor module places the address of the slave module on the bus.
- Processor module places the data on the bus.
- Processor module places the write control signal on the bus.
- Addressed slave module accepts the data from the bus.

Slave functional modules cannot exchange the data among themselves. The data exchange among the slave functional modules, if required, is always via the processor module and is totally controlled by it. Currently, there are several efficient methods for data exchange between the processor and functional modules.

8.4   Functional Subsystems

The following sections describe the construction and functioning of various subsystems of the controller and their associated functional modules.

8.4.1   Power Supply Subsystem

The controller needs power from an external source for its working. The power supply subsystem provides this through the power supply module.

8.4.1.1   Power Supply Module

Functional modules of the controller, being electronic, need power at specific voltages (+5V DC and +24V DC) for their working. To facilitate this, the power supply module takes the external AC or DC power (230/110V AC or 24/48V DC typically), converts it internally to regulated +5V DC and +24V DC (required for the operation of controller electronics), and supplies it to all the functional modules on the bus. Today's electronics do not require +24V DC and can manage with only +5V DC. Figure 8.8 illustrates the functional schematic of the power supply module.

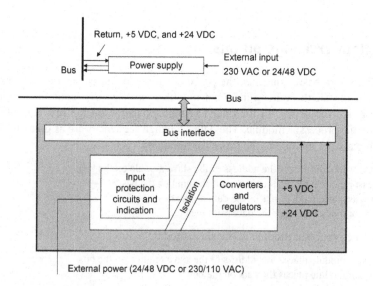

Figure 8.8 Power supply module.

8.4.2   Processor Subsystem

The processor subsystem has three modules: processor, memory, and watchdog.

8.4.2.1   Processor Module

The processor module is the heart of the controller, built around a processor, and it is responsible for the execution of the memory resident control instructions, which are based on the automation strategy. This module operates on functional modules for data acquisition and control.

As shown in Figure 8.9, the processor module has a built-in clock through which the processor derives all the necessary clocking and timing functions for synchronizing its own activities, including keeping track of the time of the day. The processor also uses the clock signals to externally synchronize the data exchange activities on the bus with the other functional modules placed on the bus. The bus interfacing electronics are located on the top portion of the processor module.

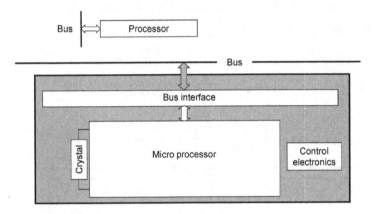

Figure 8.9 Processor module.

8.4.2.2   Memory Module

As shown in Figure 8.10, the memory module has two parts: a nonvolatile memory (ROM—read-only memory, or similar) for storing the system and automation program and a volatile memory (RAM—random access memory) for process and intermediate operational data storage.

8.4.2.3   Watchdog Module

The watchdog module, or supervision module, monitors the health of the controller and makes an announcement if it malfunctions (for both fatal and nonfatal failures). This is illustrated in Figure 8.11.

As long as the processor is healthy and functional, the controller executes the diagnostic programs in the background (whenever it is free from executing automation functions). It detects and announces any malfunctioning of functional modules. These faults, if any, are **nonfatal** in nature, as the controller is healthy except for that faulty module. The processor, being healthy, makes the announcement through the watchdog module.

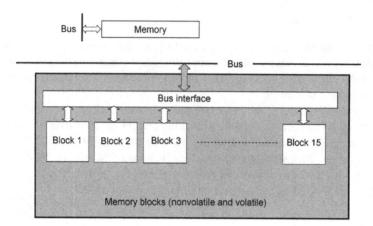

Figure 8.10 Memory module.

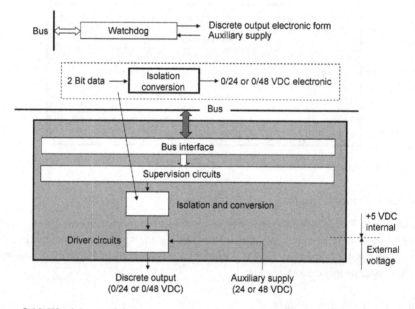

Figure 8.11 Watchdog module.

The self-supervision deals with a situation when the processor, the heart of controller, fails for some reason. This fault, being **fatal** in nature, needs to be detected and announced right when it happens. Typical fatal faults are due to the failure of the following:

- Power supply
- Processor
- Memory

Power supply failure can be due to the failure of either the external supply or the power supply module itself. The processor failure can be due to hardware failure (processor or memory) or to software failure (program going into an endless loop, or hanging). Any of these fatal faults will totally knock off the controller. The following sections discuss the detection and announcement of fatal and nonfatal faults in the controller. The watchdog **independently** announces the fatal fault when the processor is nonfunctional (either due to its failure or due to power failure).

8.4.2.3.1   Power Failure Detection and Annunciation

The power to the controller becomes unavailable due to the failure of either the external power supply or the internal power supply module. In both the cases, the effect is same—total failure of controller. Figure 8.12 illustrates the arrangement for power failure detection and its annunciation over a common audiovisual output. Here, the central detecting element is an NC relay whose contact remains open as long as the relay coil receives $+5$ V DC from the bus (meaning the controller is receiving the power). When the relay coil does not receive $+5$ V DC (meaning either the external power supply has failed or the internal power supply module has failed), the relay contact drops down to close the annunciation circuit through an auxiliary power supply.

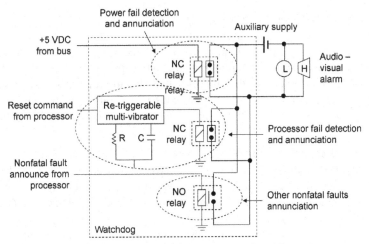

The elapse time is preset by suitable combination of R and C.

Figure 8.12 Watchdog circuits.

8.4.2.3.2   Processor Failure Detection and Annunciation

The processor can fail either due to hardware or software failure, and either case amounts to total controller failure. Figure 8.12 illustrates the arrangement for processor failure detection and annunciation over the common audiovisual output. Here, the central detecting element is an NC relay operated by a re-triggerable multi-vibrator. The multi-vibrator generates an output pulse of preset duration. This duration can be extended by re-triggering the multi-vibrator before the preset time is elapsed to remain high for the next duration. The duration starts when the trigger is applied. As long as the multi-vibrator keeps receiving the triggers periodically within the preset time durations, its output remains high and doesn't allow the relay contact to drop. If for any reason the multi-vibrator fails to receive the trigger before the end of the preset time, the relay contact drops down to drive the audiovisual output. This is illustrated in Figure 8.13.

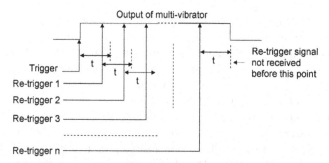

Output of multi-vibrator

Re-trigger signal not received before this point

Trigger
Re-trigger 1
Re-trigger 2
Re-trigger 3
Re-trigger n

As long as the processor keeps sending re-trigger signals periodically before the elapse of preset time "t," the output of the multi-vibrator keeps its output high. If the processor fails to send re-trigger signal before the elapse of preset time "t," the multi-vibrator drops its output to low.

Figure 8.13 Watchdog timing sequence.

The processor, if healthy, periodically issues the trigger to the multi-vibrator. Keeping the multi-vibrator output high is the highest priority task the processor performs in real-time. Whenever the processor fails due to either hardware or software fault, it cannot perform this task, which leads to the output of the multi-vibrator becoming low, relay contact getting closed, and the audiovisual alarm going high.

8.4.2.3.3 Diagnostic Error Annunciation
The processor periodically executes the diagnostic programs to check the health of all its functional modules. Whenever a fault is noticed in any functional module, the processor announces it through the audiovisual output. The diagnostic (nonfatal) error annunciation is illustrated in Figure 8.12.

The preset time is selectable using an resister-capacitor (RC) combination of the multi-vibrator. The time during which the processor is not supervised is limited to the preset time duration. Hence, the lower the preset time, the higher the self-supervision and the higher the overhead on the processor performance. On the contrary, the higher the preset time, the lower the self-supervision and the lower the overhead on the processor performance.

In this illustration, relay components are employed to simply explain the concept of the detection and annunciation of fatal and nonfatal conditions. However, the arrangement can also be with equivalent solid state circuits for higher reliability and compactness.

8.4.3   Input/Output Subsystem

In the input/output subsystem, modules of different types interface the controller with the process via the instrumentation subsystem.

8.4.3.1   Digital Input Module

The digital input module acquires the discrete electronic inputs from the instrumentation and human interface subsystems, converts them into a computer equivalent (data), and passes them on to the processor over the bus for further processing. Figure 8.14 illustrates the functional schematic of this module.

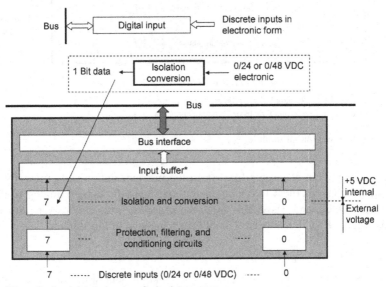

Figure 8.14 Digital input module.

8.4.3.2   Digital Output Module

The digital output module receives a computer equivalent (data) of the discrete outputs from the processor over the bus, converts it to its electronic equivalent, and sends it to the instrumentation and human interface subsystems. Figure 8.15 illustrates the functional schematic of this module.

8.4.3.3   Analog Input Module

The processor cannot understand and process analog signals unless they are converted into understandable equivalents. The analog input module acquires the

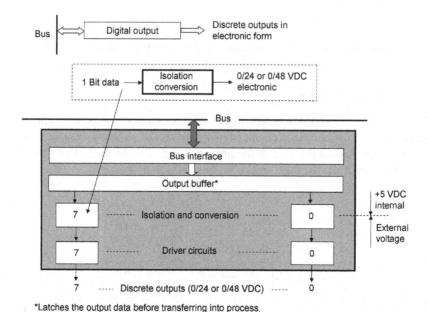

*Latches the output data before transferring into process.
Generally, a module has 8 or 16 or 32 output channels.

Figure 8.15 Digital output module.

continuous inputs in electronic form from the instrumentation and human interface subsystems, converts them into a computer equivalent through an **analog to digital converter** (**ADC**), and passes this on to the processor over the bus for further processing. Figure 8.16 illustrates the functional schematic of this module.

8.4.3.4   Analog Output Module

The analog output module receives the computer equivalent of the analog signal from the processor over the bus, converts it into a continuous electronic equivalent through a **digital to analog converter** (**DAC**), and sends it to instrumentation and human interface subsystems. Figure 8.17 illustrates the functional schematic of this module.

8.4.3.5   Pulse Input Module

The pulse input module receives the fluctuating inputs from instrumentation and human interface subsystems in electronic form from the instrumentation subsystem, counts them (serial to parallel conversion) into its computer equivalent (counter form), and passes the counter value to the processor over the bus for further processing. Figure 8.18 illustrates the functional schematic of this module.

8.4.3.6   Pulse Output Module

Similarly, the pulse output module receives the computer equivalent (counter value) of the pulses outputs (in counter form) from the processor over the bus, converts

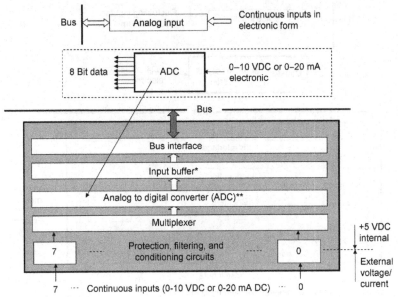

Figure 8.16 Analog input module.

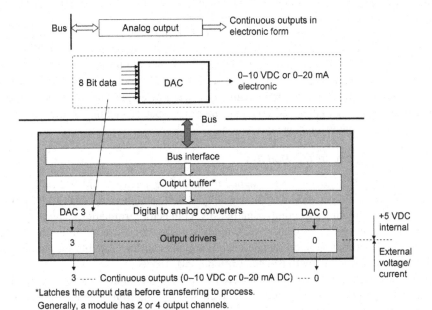

Figure 8.17 Analog output module.

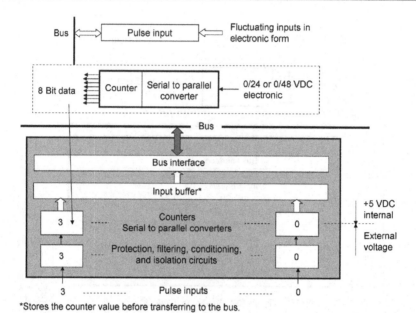

*Stores the counter value before transferring to the bus.
Generally, a module has 4, 8, or 16 input channels.

Figure 8.18 Pulse input module.

them into a pulse stream (parallel to serial conversion) in electronic form, and passes this on to the instrumentation and human interface subsystems. Figure 8.19 illustrates the functional schematic of this module.

In practice, the pulse inputs from the process (data acquisition) and the pulse outputs to the process (control execution) are of very low frequencies compared to electronic capabilities. Hence, the digital input and output modules, being quite fast in their response, can be employed for pulse input and output functions. The present digital I/O modules, which were explained earlier, are designed to accept both digital and pulse I/O. For high-speed counting, there are special modules which are commonly used in manufacturing automation.

8.4.3.7 Capacity in I/O Modules

Generally, the I/O modules are designed to have the capacity of either 2 or 4 or 8 or 16 or 32 channels per module. The capacity in I/O modules depends on the physical size of the module, its component density, connector terminations, signal types – isolated or non-isolated signals, etc. A module of the same size, with the same connector, and without isolated (single-ended) signals can have double the number of channels as the module with isolated (differential) signals.

If instrumentation devices are powered by a common source, there is no need to isolate each signal from the other. This applies to signals from human interface subsystem also. This means more channels per module and less field cabling, as shown in Figure 8.20.

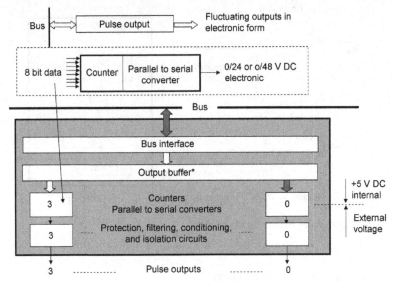

Figure 8.19 Pulse output module.

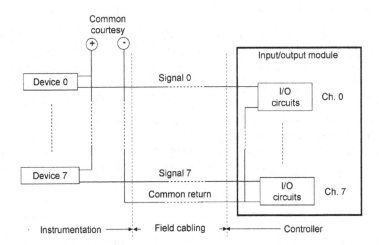

Figure 8.20 I/O modules with non-isolated signals.

If instrumentation devices are powered by different sources, there is a need to isolate each input from the other. This applies to signals from human interface subsystem also. This means fewer channels per module and more field cabling, as shown in Figure 8.21.

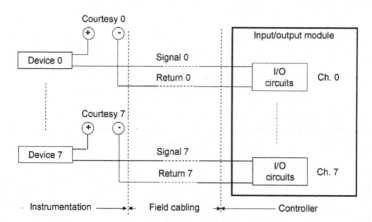

Figure 8.21 I/O modules with isolated signals.

8.4.4 Communication Subsystem

The communication subsystem has traditionally supported the following two types of interfaces (I/F) for data exchange with external compatible systems:

- *Serial I/F*: These are generally slow-speed links and are normally used for data exchange between a computer and peripherals like printers. In automation, this is also used for linking measurement, monitoring, communication equipment, etc. RS-232C is a typical example.
- *Local area network I/F*: These are typically high-speed links that can also form networks allowing multiple devices to communicate on the same link. These interfaces are also essentially serial in the way data is transmitted and exchanged, but they follow certain other standards and methodologies that allow multiple devices to simultaneously communicate over the same network at high speeds. The Ethernet is a typical example.

In the following discussion, without getting into too much complexity, we employ the commonly used terms **serial I/F** to denote the slow-speed interface and **Ethernet I/F** to denote high speed or local area network interface. Barring the internal differences, both the interfaces are serial and facilitate communication of the controller with external compatible systems.

8.4.4.1 Communication Module

The communication module is for bidirectional communication between the processor and a compatible external device or system. Typically, the communication module receives the serial pulse stream from the communication media, converts it into parallel computer data (serial to parallel conversion), and passes this on to the processor over the bus for further processing. In the reverse direction, the communication module converts the parallel computer data received from the processor over the bus into a serial pulse stream (parallel to serial conversion) and passes this on to

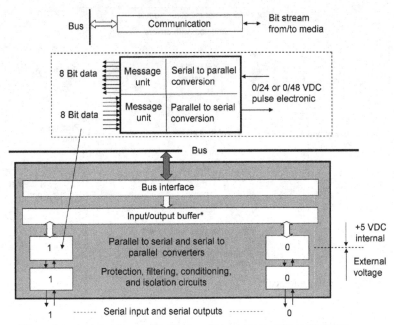

Figure 8.22 Communication module.

the communication media. This is somewhat similar to the combination of the pulse input and output module except for the meaning of the data. Over the communication media, the data transfer is always serial.

Figure 8.22 illustrates the functional schematic of this module.

8.4.4.2 Communication Cables

Figure 8.23 illustrates the communication cables employed for the two types of interfaces, which extend the signals between the communication module and the communication equipment/media—RS-232C for serial interface and Cat 5 for Ethernet interface.

8.4.5 Integrated Processor Module

Figure 8.24 illustrates the functional schematic of the integrated processor module (processor module physically and logically integrated with memory, watchdog, and communication modules over the internal bus).

Memory, watchdog, and communication modules need fast communication with the processor, so a separate **internal bus** is provided for these modules. This

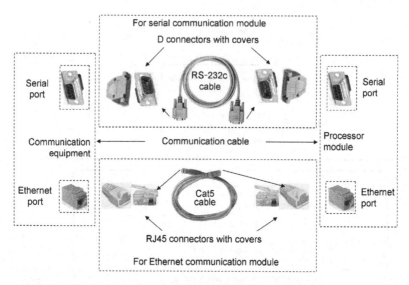

Figure 8.23 Communication cable.

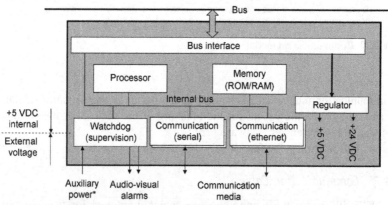

*Auxiliary power is required to drive the audio-visual alarms.

Figure 8.24 Integrated processor module.

arrangement saves the external bus for the exclusive use of the functional modules for data exchange with the processor.

An integrated module can have more than one serial and Ethernet interface, either to provide redundancy or to connect to more than one external system. Additionally, it is common to have an exclusive serial interface to connect the controller to a computer-based terminal for programming and diagnosis of the controller.

Figure 8.25 illustrates the front view of the integrated processor module with single and dual ports for each serial and Ethernet port.

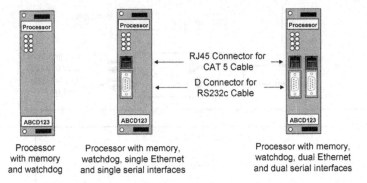

Figure 8.25 Integrated processor module—front view.

8.5 Controller Capacity Expansion

As already stated, the basic rack has a limited number of slots to hold functional modules (mainly I/O modules). Whenever the application calls for more I/O modules, extension or supplementary racks are required to house the additional modules (capacity expansion), as illustrated in Figure 8.26. The supplementary racks only physically extend the bus controlled by the processor in the main rack. Bus amplifier/driver modules are used to extend the bus. The number of racks and functional modules depends on the capability of the processor module to handle them.

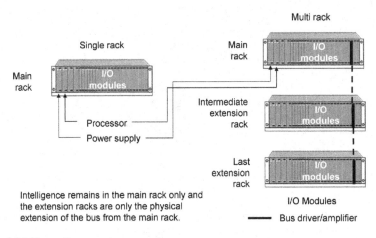

Figure 8.26 Controller capacity expansion.

8.5.1 Bus Extension (Parallel) Module

Figure 8.27 illustrates the schematic of the bus extension (parallel) module. This module amplifies (repeats) the signals and drives them in both directions, in addition to supporting the rack in which the module is located.

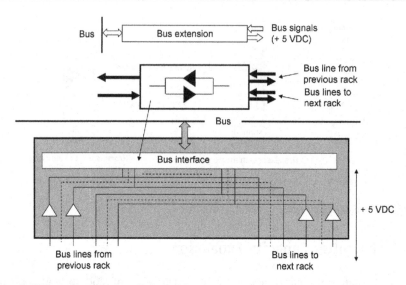

Figure 8.27 Bus extension (parallel) module.

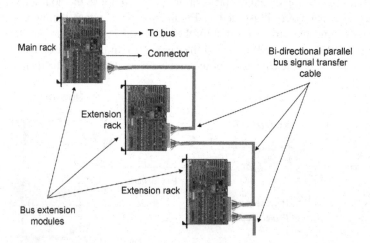

Figure 8.28 System cable for bus extension.

Figure 8.28 illustrates the prefabricated system cables for bus extension and their connections among the busses in main and extension racks.

8.5.2 Bus Extension (Serial) Module

In the previous example, all the bus signals are extended through parallel signal transmission (in both directions) from one rack to the other with a bus extension (amplifier/repeater) module. Figure 8.29 illustrates the bus extension through serial

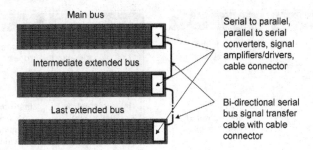

Figure 8.29 Bus extension (serial).

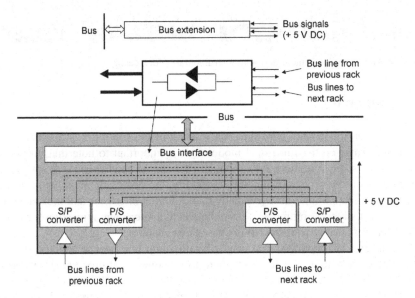

Figure 8.30 Bus extension module (serial).

signal transmission in both directions (serial to parallel and parallel to serial conversions with signal amplifiers/repeaters), which are built into the bus itself. Figure 8.30 illustrates the bus extension (serial) module. This arrangement facilitates remote placement of the extension racks, as serial signals can be driven over longer distances.

8.6   Integrated Controller

Figure 8.31 illustrates the completely integrated controller with instrumentation devices. This arrangement can have either an individual power supply module in each rack or a common power supply powering all the racks within the cabinet.

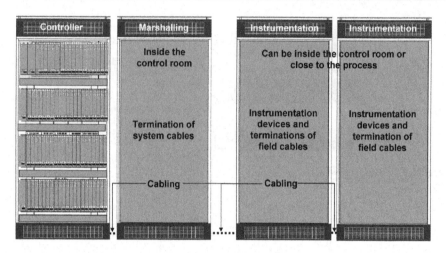

Figure 8.31 Controller—integrated with instrumentation devices.

8.7 Summary

In this chapter, the general hardware construction of the controller is discussed with special emphasis on the functional modules. It is important to note that, in present day controllers, the communication modules (both serial and Ethernet) are physically integrated with the processor module. Generally, the processor modules have multiple communication interfaces (typically dual Ethernet interfaces and multiple serial interfaces) to support communication redundancy and/or communication with multiple external systems, as discussed in subsequent chapters.

Present day I/O modules are designed with some intelligence (microprocessor/ microcontroller-based), which allows them to do some local processing and operations to reduce the routine data transfer load on the bus and the computing load on the processor. This improves the overall performance of the controller. These issues are discussed in detail in Chapter 16.

9 Software Structure of Controller

9.1 Introduction

In Chapter 8, we discussed hardware and engineering aspects of a general-purpose controller. Over the years, a lot of advances have taken place in hardware technologies (mainly in terms of electronics and communication). However, as explained in Appendix C, the hardware interfaces have remained virtually the same while control has become more powerful due to large memory and increased processor speeds. Apart from compactness and reliability, memory and speed constraints, present in earlier systems, are no longer valid today. Because of this, software has overtaken hardware in performing many tasks. In fact, software is eliminating the need for certain aspects of hardware interface electronics, and software is taking care of the complete operation and control of industrial processes. Software-based systems are more flexible, allowing for modifications, if required. They also provide maximum facilities, leaving the user only to customize the automation for particular processes. To sum up, *software in today's automation system does almost everything* in operation, monitoring, and control of industrial processes. In this chapter, the software architecture of the controller is discussed.

9.2 Types of Software Systems

The software systems can be broadly classified as non-real-time and real-time. **Real-time** systems take certain actions that follow the clock or work according to the **time-of-day**.

The general-purpose system is a **non-real-time** system. Within this type of system, there are no deadlines for the response to an event. In a real-time system, there is a deadline for the response to an event. Further, the real-time systems can be soft or hard. In **soft real-time systems**, not meeting the deadline can have undesirable effects, but they do not lead to dangerous situations since the response will eventually happen some time later. A real-time passenger reservation system is a typical example. On the contrary, in **hard real-time systems**, not meeting a deadline can lead to dangerous effects. Real-time monitoring and control of a nuclear plant is a typical example. Automation systems are hard real-time systems.

The controller in an automation system is a hard real-time system, and its significant differences from a non-real-time system are shown in Figure 9.1.

The following section discusses the two types of systems, non-real-time and real-time, and focuses on their differences in terms of usage.

Overview of Industrial Process Automation. DOI: 10.1016/B978-0-12-415779-8.00009-7

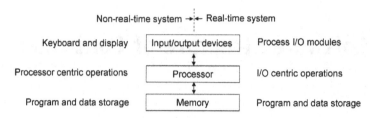

Figure 9.1 Real-time vs. non-real-time architecture.

9.2.1 Non-Real-Time System

Table 9.1 illustrates the subsystems and their functions in a non-real-time system. The hardware components are relatively less reliable, as electromechanical devices such as hard drives are more prone to failure. This is not an issue anymore since present computers support large semiconductor primary and secondary memory (flash drives) in place of hard disk drives. The operating software here is non-real-time.

Table 9.1 Non-Real-Time System

Subsystems	Functions	Examples
Processor	Execution of the stored program	Computer, mini-computer
Memory	Storage of program and data	RAM, HDD, CD, DVD, flash
Inputs/outputs	Operator interaction	Keyboard, VDU display, printer
Communication	Data exchange with external systems	Serial, LAN, WAN, Internet
Operating software	General-purpose data processing	Windows, Unix, Linux
Execution of program	Sequential and time shared	Regular data processing jobs

9.2.2 Real-Time System

Automation systems have to be hard real-time to meet the real-time needs of the industrial process. For this, the whole system needs be very time responsive to ensure results when required. For example, accidental radiation in a nuclear reactor must get immediate response from the controller so that emergency actions may be taken automatically and with minimum delay to protect people from nuclear hazards.

Table 9.2 illustrates the subsystems and their functions in a controller. The hardware components need to be maintenance-free, as maximum reliability and availability is called for in the function of automation systems. Also, the operating software must meet the time constraints of various operations to produce responses and results within the allotted time.

The objective of industrial process automation systems is to enable the process plant to produce desired results. This calls for the execution of automation programs in real-time. An automation program is a set of interlinked software subroutines or tasks. A **task** is a set of instructions (code) that is executed by the processor under the control of the scheduler. In computer language, a subroutine is an executable

Table 9.2 Processor Subsystem Functions

Subsystems	Functions	Examples
Processor	Execution of the stored program	Microprocessors
Memory	Storage of program and data	RAM, ROM, PROM, E^2PROM, flash
Inputs/outputs	Process interface	I/O modules
	Operator interaction	I/O modules for mimic control panel
Communication	Data exchange with external systems	Serial, LAN, WAN, Internet
Operating software	Real-time data processing	VxWorks, Windows CE, RTLinux
Program execution	Priority-based	Real-time tasks

code. A **call subroutine** instruction in a program transfers program control to the subroutine. The subroutine, on completion of its execution, returns to the called program through a **return from subroutine** instruction.

Since a task is a subroutine, it can be called by any program, when required, to execute a specific function and return to that program on completion. In real-time context, all tasks have an allotted time for their execution after initiation. When a task becomes due for execution, it must be given the required resources so that it is executed within its allotted time. Non-real-time tasks (say, servicing of printer) can be executed when there are no real-time tasks pending for execution. Real-time operating systems (RTOSs) facilitate such requirements.

9.3   Software Structure of Controller

Figure 9.2 illustrates the general software structure of the controller. Each lower level layer has the interface to interact with its next higher level. Here, except the innermost layer, all the layers are software.

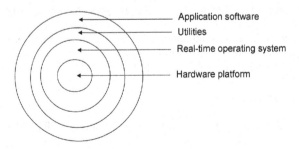

Figure 9.2 Software structure in controller.

The following sections explain the functions of the layers in a software structure.

9.3.1   Hardware Platform

The hardware platform layer basically consists of hardware resources, such as a processor subsystem (processor, memory, and watchdog), communication subsystem (serial and Ethernet interface), and input/output (I/O) subsystem (all types of input and output) along with its interfaces to the operating system.

9.3.2   Real-Time Operating System

A real-time operating system is designed for real-time operations meeting response time predictability and deterministic response. The main criteria are predictability and determinism, not speed. It is also possible to have both soft and hard real-time features in the same system. Excessive overhead in operating system software can affect its response time and performance. Unlike general-purpose operating systems, RTOS adds only a very small overhead in microseconds while the response requirement in automation systems is in milliseconds or above. This leaves a good portion of computing power for execution of non-real-time tasks.

The kernel of RTOS, in its basic form, is a memory resident software that takes responsibility for the overall management of the real-time system by responding to time- and event-controlled tasks. The main functions are as follows:

- *Resource management*: Sharing of resources by the competing tasks as per their execution schedules. This means that the tasks get the required resources allocated to them whenever they are needed.
- *Task management*: Creation of the task and its activation, running, blocking, resumption, de-activation
- *Task scheduling*: Scheduling multiple tasks either on a cyclic/programmed basis or on a noncyclic/pre-emptive basis with strict adherence to the schedule, leading to predictable or deterministic results
- *I/O management*: Executing service tasks either on a programmed basis or on a priority basis
- *Memory management*: Allocating and de-allocating the memory for tasks
- *Inter-task communication and synchronization*: Sending messages from one task to another task for their synchronization. This becomes necessary during the parallel processing of tasks, especially when execution of some tasks is dependent upon the completion of other tasks.
- *Interrupt management*: Handling multilevel priority-based interrupts (see Section 9.4.3.1 for more on program interrupts)
- *Time management*: Keeping real-time and time-of-day, facilitating time-based execution of tasks and monitoring of elapsed time of tasks after their initiation. Timers have the highest level of interrupts and keep track of real-time.

9.3.3   Utility Software

Utility software is the standard and commonly used application programs.

9.3.4   Application Software

Application software performs automation functions specifically customized for the industrial process.

Real-time operating systems provide facilities with links to all the resources (hardware and software). Thus, the application programmer only needs to know the links to the real-time operating system to access all the available resources.

9.4 Scheduling of Tasks

As already mentioned, the application software is divided into several tasks for performing control functions in a predetermined manner. This is called scheduling of tasks. Each task remains in one of the following three states:

- *Running*: Currently under execution
- *Ready*: Waiting to be executed
- *Waiting*: Waiting for an external event, so not ready for execution

The following sections explain how the tasks are scheduled for execution to meet different requirements.

9.4.1 Sequential Scheduling

In sequential scheduling, as illustrated in Figure 9.3, all eligible tasks are of equal priority and are executed one after the other in a cycle. There is no time criticality for any task in order to produce the results.

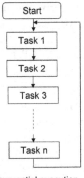

Sequential execution

Figure 9.3 Sequential execution of tasks.

9.4.2 Sequential Scheduling with Time-Slice

In sequential execution, a task with a long execution time can excessively delay other tasks in the cycle from getting executed. This is not always desirable. One way to partially overcome this problem is to go for sequential scheduling with time-slice, which involves allotting a fixed time-slice in sequence to all eligible tasks for their execution. In this method, each task with a fixed time-slice is partially executed, and it paves the way for the next task in the queue to begin in next time-slice. The process goes on until all eligible tasks are fully executed.

9.4.3   Real-Time Scheduling

In real-time scheduling, the tasks are assigned to events that have different priorities, and each event is serviced by executing its associated task. The scheduler services the events based on their priority. Each event can interrupt the processor and get its task serviced on priority, as explained in Section 9.4.3.1.

9.4.3.1   Program Interrupt

Program interrupt, a hardware mechanism that allows events to get serviced, can interrupt the execution of an ongoing task. An event can interrupt provided it is allowed by the system to interrupt. Whenever the processor receives an interrupt, it suspends execution of the ongoing task and allocates resources to the event that has interrupted, provided the interrupting event is of higher priority than the one currently under execution. To facilitate this, the processor checks for interrupts at the end of every instruction before moving on to the next instruction. If any interrupt is pending, the processor automatically branches off to interrupt service routine (like branching off to a subroutine) after saving the status of the interrupted program. Saving the status (mainly the address to which the processor should return) of the interrupted program is essential for returning to interrupted program after servicing the request. This entire sequence is similar to calling a subroutine and returning from a subroutine. The only difference is that a subroutine call is programmed, while the interrupt call is random from the hardware or software.

Like subroutines, servicing of interrupts can be nested. This means that at any point of time, events with higher priority can interrupt the servicing of events with lower priority. Each time a processor receives an interrupt, the interrupted program status is saved in a first-in/last-out stack and the restoration of the interrupted program takes place in reverse sequence. The RTOS manages the interrupt hierarchy.

Typical examples of hardware-generated events that can interrupt are as follows:

- Diagnostic error in functional modules
- Change of state in digital input module
- Counter becoming full in pulse input module
- Counter becoming empty in pulse output module
- Message-in buffer becoming full in communication module
- Message-out buffer becoming empty in communication module

Typical examples of software interrupt are as follows:

- Timer or time-of-day interrupts for servicing time-based events
- Programming errors, such as division by zero

Generally, all modes of task scheduling are employed in today's RTOS for different requirements.

9.4.3.2   Task Execution

The interrupt manger in the RTOS handles multilevel interrupts and their nesting. Generally, all events in the queue get serviced in time, as external devices and systems need attention in milliseconds or seconds, while the processor response is in

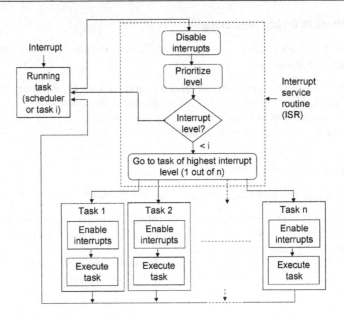

Figure 9.4 Real-time execution of tasks.

microseconds. Figure 9.4 illustrates nesting of real-time task scheduling and task execution. This is an example with a single interrupt line over which many devices interrupt the processor. Here also, the last instruction in each task is *return from subroutine* so that the task returns to the scheduler or interrupted task after completion of its execution.

9.5   Scheduling of Tasks in Automation Systems

The controller employs the RTOS and works with real-time task scheduling for the execution of automation tasks. Some basic controller tasks are discussed in the following sections.

9.5.1   Process Data Acquisition

The following are real-time activities in process data acquisition:

- The most important is the acquisition of changes in digital input states when they occur (e.g., the occurrence/disappearance of a process alarm). There is a possibility of losing this information if another change takes place before the current change is acquired.
- Acquisition of the values of analog inputs more frequently, as their values may vary fast (e.g., a continuous varying of a process value). There is a possibility of missing the continuity of information if a substantial variation takes place before the current value is acquired.

9.5.2   Process Data Monitoring

In process data monitoring, recognize the following events as early as possible to initiate the intended actions:

- Change in the states of raw digital inputs (e.g., the occurrence of an alarm) or the derived inputs as a function of basic inputs (e.g., generation of a group alarm)
- Variation in the values of raw analog inputs (e.g., the occurrence of limit violations), derived inputs, or derived inputs as a function of basic inputs

9.5.3   Process Control

The following are the real-time activities in process control:

- Computing and sending variable continuous control command (analog output) as per the required periodicity or when required to the final control element to regulate the process performance (e.g., maintaining accurate flow)
- Sending discrete control command (digital output) to the final control device without delay when it becomes necessary (e.g., the emergency shutdown command)

9.6   Memory Organization

System generation is an offline process for configuring and customizing the hardware and the software for a specific application to create a **run-time system**. During system generation time, the programmer configures the hardware and software, creates a run-time system, and loads it into program memory.

Memory in the real-time system for program and data is typically organized during **run-time**. The total address space in the system is shared by the program and data. Figure 9.5 illustrates the typical run-time memory mapping.

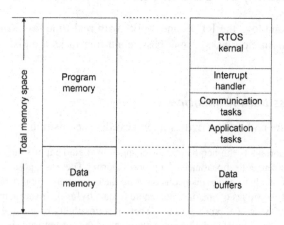

Figure 9.5 Run-time memory mapping.

9.7 Summary

In this chapter, we discussed the basics of software structure in the controller with special emphasis on real-time operating systems and their functions, such as resource management, task management, task scheduling, I/O management, memory management, inter-task communication and synchronization, and time management. Different types of task scheduling—sequential, sequential with time-slice, priority- or event-based, and real-time—were discussed with specific reference to the automation system.

10 Programming of Controller

10.1 Introduction

Unless programmed for specific automation functions, the controller cannot perform the desired functions. Generally, the automation strategy is programmed by automation engineers. The job of an automation engineer is to configure or customize the controller with the specific automation program to meet the requirements of the process, in addition to configuring the controller with the required inputs and outputs. Appendixes B, C, D, and E describe basic programming in lower-level languages (machine and assembly). This chapter gives an introduction to the programming of the controller in higher-level languages with the tools and procedures provided by controller vendors for coding the programs.

10.2 Higher-level Programming

Coding the automation program in assembly-level language makes the program executable only on the specific machine or platform for which the program has been coded. Hence, programming of the controller is done in higher-level languages that are easily understood by the programmers and make the program **portable** for use on other controllers and platforms with minimum adaptation. The compiler program, developed by the platform vendor, converts this higher-level language program into its machine executable equivalent and downloads it to the controller (target machine).

The controller is programmed for automation functions by using the following higher-level languages, which are covered by the **IEC 61131-3 standard**:

- Ladder diagram (LD)
- Function block diagram (FBD)
- Structured text (ST)
- Instruction list (IL)
- Sequential function chart (SFC)

Figure 10.1 illustrates the structure of these higher-level languages, which are generally supported by all the vendors. The IEC 61131-3 standard allows mixing of automation programs written in all of these higher-level languages.

The most popular languages used by the automation engineers are LD and FBD. They are graphic-based and were created to make program development and maintenance easier for automation and process engineers.

Overview of Industrial Process Automation. DOI: 10.1016/B978-0-12-415779-8.00010-3

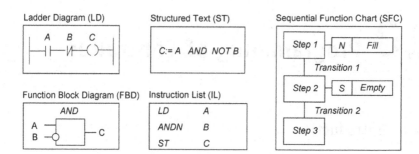

Figure 10.1 Higher-level programming languages.

ST is a text-based language similar to C/Pascal, and IL is also text-based similar to assembly language. Both these languages are used by application software developers. SFC is graphic-based with a time- and event-based interface between controller and user during the program development, startup, and troubleshooting. In the following sections, we discuss the salient features of LD and FBD.

10.2.1 Ladder Diagram

The ladder diagram (LD) originated in the graphical representation of electrical control systems using relays (relay-based logic). It is mostly used for discrete automation and is ideal for sequential control with interlocks. The specifics of LD are as follows:

- Based on the schemes/circuit diagrams of relay logic programming, as discussed in Appendix A
- Graphical representation for the programming elements

The name ladder diagram is derived from the program's resemblance to a ladder with two vertical rails and a series of horizontal rungs between them. The rails are called power rails in the ladder diagram. Figure 10.2 illustrates a typical ladder diagram and its conventions.

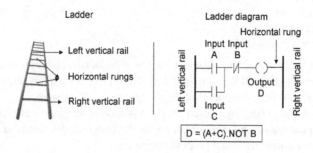

Figure 10.2 Ladder diagram.

Table 10.1 LD—Basic Instructions

Symbol	Function	Operation
A ─┤├─ B	NO—Normally open	Normally, O/P B is OFF. O/P B becomes ON when the I/P A is ON.
A ─┤/├─ B	NC—Normally closed	Normally, O/P B is ON. O/P B becomes OFF when I/P A is ON.
─(A)─	Output—Multiple output can be saved in multiple parallel	Normally, O/P A is OFF. O/P A becomes ON when all I/Ps in the rung are ON. Generally, this is the last instruction in the rung. O/P can be stored as a flag in memory for subsequent use.
A ─┤In TON├─ C B ─┤Preset│	TON—On delay timer	Normally, O/P C is OFF. When I/P A is enabled (becoming ON), O/P C becomes ON only after the predetermined time B has elapsed.
A ─┤In TOF├─ C B ─┤Preset│	TOF—Off delay timer	Normally, O/P C is ON. When I/P A is enabled (becoming ON), O/P C becomes OFF only after the predetermined time B has elapsed.
A ─┤In O├─ D B ──┤Reset│ C ──┤Preset│	CTU—Up counter	Initially, O/P D is OFF. When each time I/P A transitions from OFF to ON, the counter increments. When the counter reaches the preset value C, O/P D goes from OFF to ON. I/P B resets the preset value.
A ──┤In O├─ D B ──┤Reset│ C ──┤Preset│	CTD—Down counter	Initially, O/P D is OFF. When each time I/P A transitions from OFF to ON, the counter decrements from the preset value C. When the counter becomes 0, O/P C goes from OFF to ON. I/P B resets the preset value.

Table 10.1 illustrates the basic LD programming symbols/instructions for sequential control with interlocks. Today, LD supports a lot more programming instructions, including mathematical functions required in continuous process automation.

Table 10.2 illustrates some simple logical functions programmed with bit logic instructions of LD.

Table 10.3 illustrates some simple application examples programmed with bit logic instructions in LD.

Table 10.2 Logic Instructions

Symbol	Function	Operation
I/P A ┤├──────(B)── O/P	Repeater/buffer	O/P B is enabled (becomes ON) when I/P A is enabled (becoming ON). B = A
I/P A ┤/├──────(B)── O/P	Inverter/NOT	O/P B is disabled (becomes OFF) when I/P A is enabled (becoming ON). B = NOT A
I/P B ┤├───┤├──(C)── I/P A O/P	2 Input AND	O/P C is enabled (becomes ON) only if both I/P A and I/P B are enabled (becoming ON). C = A · B
I/P A ┤├─┐ ├──────(C)── ┤├─┘ O/P I/P B	2 Input OR	O/P C is enabled (becomes ON) if either I/P A or I/P B or both are enabled (becoming ON). C = A + B

Table 10.3 Application Examples

Problem 1: O/P latching function

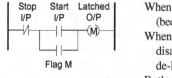

When start I/P switch is momentarily pressed, O/P M gets enabled (becomes ON), sets its flag M, and gets latched (stays ON).

When stop I/P switch is momentarily pressed, O/P M gets disabled (becomes OFF), resets its flag M, and gets de-latched (stays OFF).

Both start and stop I/P switches are momentary types in action.

Problem 2: Car door open alarm generation function

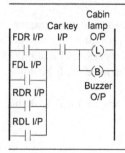

Any one or more doors getting opened (I/Ps becoming ON) and car key I/P in enabled state (ON) activate cabin lamp O/P and buzzer O/P (becoming ON).

Table 10.3 Application Examples (Continued)

Problem 3: Parking full indication and closing the gate

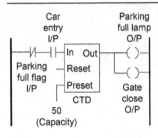

Parking lot can hold 50 cars. The counter is preset to 50.
 Entry of every car decrements the counter.

As long as the number of cars in the parking lot is less
 than 50, the counter O/P is OFF.

Parking full flag, being NC, keeps the circuit closed and
 car entry input decrements the counter every time a car
 enters.

When the counter becomes zero, parking full O/P becomes
 ON, gate close output also becomes ON closing the
 gate and making the parking full flag input open, thus
 opening the circuit.

Problem 4: Lift door open operation

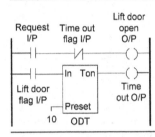

The time required to open the lift door upon pressing the
 passenger request button is 10 seconds. The ON delay
 counter is programmed for 10 seconds.

Upon passenger registering his request, the circuit gets
 closed, door motor starts, and timer also starts.

Upon the timer completing 10 seconds, the timer
 holding switch opens and the circuit opens, thus
 stopping the motor.

10.2.2 Function Block Diagram

The function block diagram (FBD), also a graphical representation of various mathematical and logical functions, is primarily developed for programming continuous process automation, even though it also supports discrete process automation.

The function block diagram, like the ladder diagram, is based on the following:

- Graphical representations of programming elements, which are modular, repeatable, and reusable in different parts of the program.
- The function block represents the functional relation between the inputs and outputs.
- The program is constructed using function blocks that are connected together to define the data exchange.
- In the programs, the values flow from the inputs to the outputs, through the function blocks.
- The primary concept behind FBD is data flow. The connecting lines have data types that must be compatible on both ends.
- Supports programming of binary numbers (digits 0 and 1) to deal with bit logic.
- Supports programming of integers (single and double) and real numbers.

Figure 10.3 illustrates the structure of a function block and a function block diagram.

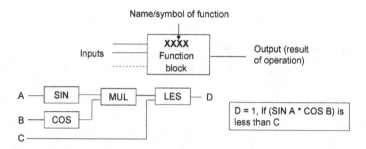

Figure 10.3 Function and function block diagram.

The following section explains some of the basic and commonly used FBD instructions.

Table 10.4 illustrates the bit logic instructions of FBD. Bit logic instructions interpret signal states of 1 (YES) and 0 (NO) and combine them according to Boolean logic.

Table 10.4 FBD—Logic Instructions

	Insert Function	
	NOT function I/P negate, O/P negate	FBD does not have a function for NOT. Inputs and outputs of function blocks can be inverted by introducing a small circle at the point of intersection.
A — [>=1] — C B —	2 I/P OR function	C becomes 1 when either A or B or both are 1. C = A + B
A — [&] — C B —	2 I/P AND function	C becomes 1 when both A and B are 1. C = A · B
A — [=]	Assign function	This is similar to O/P function in LD—produces the result of a logic operation, 1 if the conditions are satisfied, 0 if the conditions are not satisfied.

Table 10.5 illustrates the compare instructions for integers (I). Here, the inputs are integers or double integers or real (floating point), and the output is binary. If the comparison is true, the output of the function is 1 (YES). Otherwise, it is 0 (NO).

Table 10.5 FBD—Compare Instructions

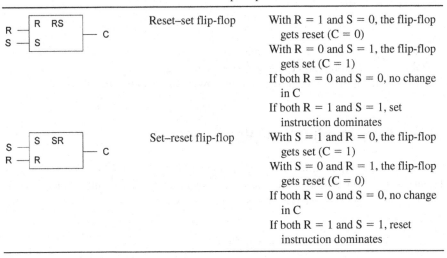

A — CMP ==I — C / B	Integer compare	If A and B are equal, then C = 1.
A — CMP >I — C / B	Integer greater than	If A is greater than B, then C = 1.
A — CMP <I — C / B	Integer less than	If A is less than B, then C = 1.
A — CMP <>I — C / B	Integer not equal	If A is not equal to B, then C = 1.
A — CMP >=I — C / B	Integer greater than or equal	If A is greater than or equal to B, then C = 1.
A — CMP <=I — C / B	Integer less than or equal	If A is less than or equal to B, then C = 1.

The above function blocks are for integer operands. Replacing I by D and by R refers to similar instructions for double integer and real operands. Output C is always binary.

Table 10.6 illustrates the basic flip-flop instructions.

Table 10.6 FBD—Flip-Flop Instructions

R — R RS — C / S — S	Reset–set flip-flop	With R = 1 and S = 0, the flip-flop gets reset (C = 0) With R = 0 and S = 1, the flip-flop gets set (C = 1) If both R = 0 and S = 0, no change in C If both R = 1 and S = 1, set instruction dominates
S — S SR — C / R — R	Set–reset flip-flop	With S = 1 and R = 0, the flip-flop gets set (C = 1) With S = 0 and R = 1, the flip-flop gets reset (C = 0) If both R = 0 and S = 0, no change in C If both R = 1 and S = 1, reset instruction dominates

Table 10.7 illustrates counter (UP and DN) instructions. In an UP counter, the rising edge (change in signal state from 0 to 1) at input increments the counter and produces an output of 1 when the counter reaches the preset value. Similarly, in a DN counter, the rising edge (change in signal state from 0 to 1) at input decrements the counter from the preset value and produces an output 1 when the counter value reaches zero.

Table 10.7 FBD—Counter Instructions

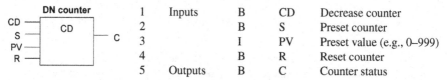

	1	Inputs	B	CU	Increment counter
	2		B	S	Preset counter
	3		I	PV	Preset value (e.g., 0–999)
	4		B	R	Reset counter
	5	Outputs	B	C	Counter status

Counter is preset to PV with change from 0 to 1 in input S.
Counter is reset to 0 with change from 0 to 1 in input R.
Change in input CU from 0 to 1 increases the counter unless the counter value is already 999.
Output C becomes 1 if the counter value becomes equal to PV.

	1	Inputs	B	CD	Decrease counter
	2		B	S	Preset counter
	3		I	PV	Preset value (e.g., 0–999)
	4		B	R	Reset counter
	5	Outputs	B	C	Counter status

Counter is preset to PV with change from 0 to 1 in input S.
Counter is reset to 0 with change from 0 to 1 in input R.
Change in input CD from 0 to 1 decreases the counter unless the counter value is already 0.
Output C is 1 if the counter value becomes 0.

Notes: B—Boolean; I—integer

Table 10.8 illustrates the basic timer instructions.
Table 10.9 illustrates some basic arithmetic instructions.
Table 10.10 illustrates the programs for some simple applications.

10.3   Programming Examples

The following sections illustrate programming examples with LD and FBD for different types of automation strategies, such as sequential control with interlocks, continuous control, and two-step control with dead-band.

10.3.1   Sequential Control with Interlocks

Figure 10.4 illustrates the water heating process with its instrumentation subsystems (discrete input and discrete output devices).

Figure 10.5 illustrates the configuration of the controller for automation of the water heating process (sequential control with interlocks).

Table 10.8 FBD—Timer Instructions

On delay timer		Inputs	B	S	Start
	2		Time	TV	Preset time (e.g., 0–999)
	3		B	R	Rest input
	4	Outputs	B	C	Timer status

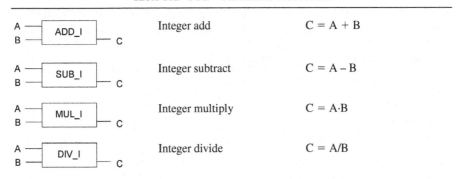

If input S changes from 0 to 1 (rising edge of start), timer starts.
If the specified time TV elapses and the state of input S is still 1, output C is 1.
If input S changes from 1 to 0, the timer stops and output C becomes 0.
If input R changes from 0 to 1 while the timer is running, the timer is restarted.

Off delay timer		Inputs	B	S	Start
	2		Time	TV	Preset time (e.g., 0–999)
	3		B	R	Rest input
	4	Outputs	B	C	Timer status

If input S changes from 1 to 0 (trailing edge of start), timer starts.
Output C is 1 when input S is 1 or the timer is running.
Output C becomes 0 after the elapsed time.

Notes: B—Boolean; Time—controller time

Table 10.9 FBD—Arithmetic Instructions

ADD_I	Integer add	$C = A + B$
SUB_I	Integer subtract	$C = A - B$
MUL_I	Integer multiply	$C = A \cdot B$
DIV_I	Integer divide	$C = A/B$

The above function blocks are for integer operands. Similar instructions can be applied to double integers and real operands by replacing I by D or R. Output C is always an integer or double integer or real (as per the inputs).

Table 10.10 FBD—Application Examples

Problem 1: Start/stop of motor

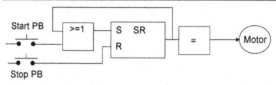

PB: Push button switch – momentary

When Start PB is pressed, input to S becomes 1. With input to R being 0, the flip-flop is set to start the motor. Feedback from FF holds the output of FF.

When Stop PB is pressed, input to R becomes 1. With input to S being 1, the flip-flop is reset to stop the motor.

Both the PBs are momentary push button switches.

Problem 2: Car cabin light/buzzer

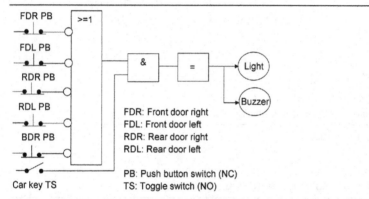

With car key TS (toggle switch) turned on and with one or more doors of FDR, FDL, RDR, RDL, PB (momentary switch) is getting opened, which activates the light and the buzzer.

Problem 3: Parking lot full/not full indication

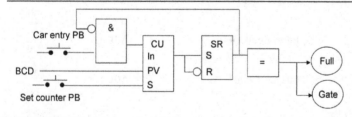

PB: Push button (NO)

When the parking lot is not full, entry of each car increments the counter. When the counter becomes full, parking lot full indication comes, gate gets closed, and further increment of the counter are blocked.

Table 10.10 FBD—Application Examples (Continued)

Problem 4: Lift door operation

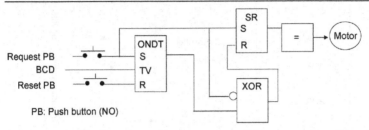

PB: Push button (NO)

On request, the flip-flop gets set, the door motor and the counter start. When the time is up, the flip-flop gets reset and stops the motor.

Problem 5: Two-step control of air conditioner

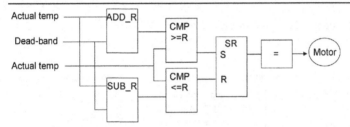

SR flip-flop is set when the actual temperature is ≥ (reference temperature + dead-band) starting the compressor motor.

SR flip-flop is reset when the actual temperature is ≤ (reference temperature – dead-band) stopping the compressor motor.

Problem 6: Speed control of motor

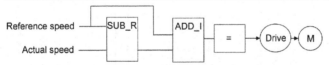

Drive receives the control signal continuously to produce variable voltage (for DC motor) or variable frequency (for AC motor) to reduce the error between the reference and actual speeds.

Table 10.11 illustrates the allocation of I/O channels to various inputs and outputs.

Table 10.12 illustrates the allocation of memory locations used for specific purposes.

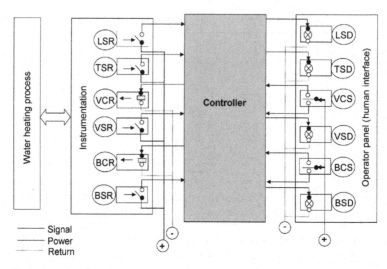

Figure 10.4 Water heating process automation (discrete).

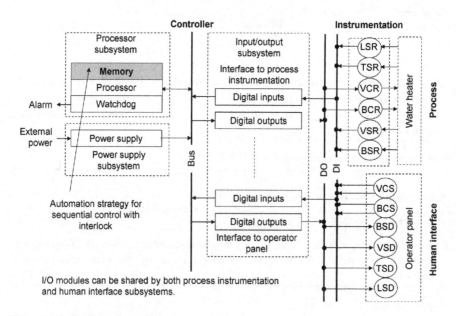

Figure 10.5 Controller configuration.

Table 10.11 Allocation of I/O Channels to Process Signals

From/To	I/O Channel	Instrumentation Device		Signal
Digital input module (8 channels—DI0–DI7)				
From process	DI0	LSR	Level switch relay	Actual level status
	DI1	TSR	Temp switch relay	Actual temp status
	DI2	VSR	Valve status relay	Actual valve status
	DI3	BSR	Breaker status relay	Actual breaker status
From operator	DI4	VCS	Valve control switch—open	Open valve command
panel	DI5		Valve control switch—close	Close valve command
	DI6	BCR	Breaker control relay—close	Close breaker command
	DI7		Breaker control relay—open	Open breaker
Digital output module (8 channels—DO0–DO7)				
To operator panel	DO0	LSD	Level status indication lamp	Actual level status
	DO1	TSD	Temp status indication lamp	Actual temp status
	DO2	VSD	Valve status indication lamp	Actual valve status
	DO3	BSD	Breaker status indication lamp	Actual breaker status
To process	DO4	VCR	Valve control relay	Open/close valve
	DO5	BCR	Breaker control relay	Close/open breaker
Not used	D06, D07			

Table 10.12 Allocation of Memory Locations for Special Requirements

Name	Function	Purpose
FVO	Flags	To remember issue of momentary valve open command
FVC		To remember issue of momentary valve close command
FBO		To remember issue of momentary breaker open command
FBC		To remember issue of momentary breaker close command
XXX		To store intermediate results

Figure 10.6 illustrates the flow chart of the automation strategy.

Figures 10.7 and 10.8 illustrate the allocation of variables (global and program) and the coded program in a ladder diagram.

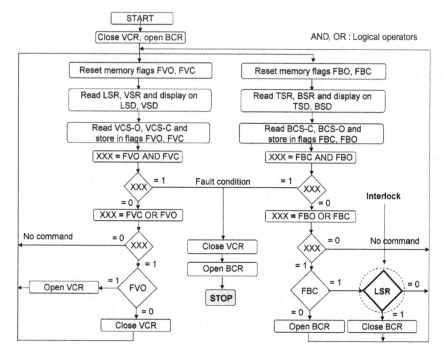

Figure 10.6 Sequential control with interlocks—flow chart.

```
VAR_GLOBAL
    //inputs
    LSR,TSR,VSR,BSR: BOOL;
    VCS_O,VCS_C,BCS_O,BCS_C: BOOL;
    INIT:BOOL: = TRUE;
    //outputs
    LSD,TSD,VSD, BSD: BOOL;
    VCR,BCR: BOOL
END_VAR
```

```
VAR_PROGRAM
    XXX: REAL
    FVO,FVC: BOOL;
    FBO,FBC: BOOL;
    XXX1,XXX2,XXX3,XXX4: BOOL;
END_VAR
```

Figure 10.7 Allocation of variables.

The LD program in Fig. 10.8 was coded and tested using CoDeSys software.[1]

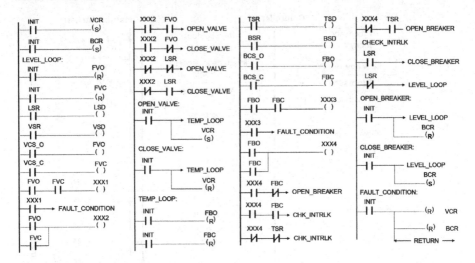

Figure 10.8 LD program.

10.3.2 Loop Control

Figure 10.9 illustrates a water heating process with its instrumentation subsystems (continuous input and continuous output devices).

Figure 10.10 illustrates the configuration of the controller for the automation of a water heating process (loop control).

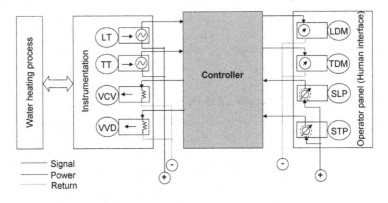

Figure 10.9 Water heater with automation (continuous).

Table 10.13 illustrates the allocation of I/O channels to various inputs and outputs.
Table 10.14 illustrates the allocation of memory locations used for specific purposes.
Figure 10.11 illustrates a flow chart of the automation strategy.

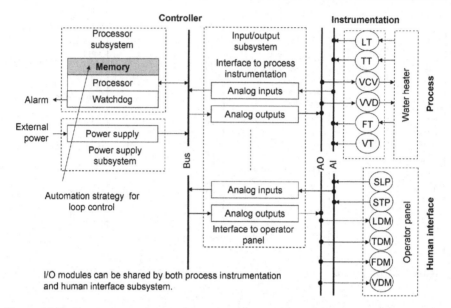

Figure 10.10 Controller configuration.

Table 10.13 I/O Channel Allocation to Process Parameters

From/To	I/O Channel		Instrumentation Device	Signal
Analog input module (8 channels: AI0–AI7)				
From process	AI0	LT	Level transmitter	Actual water level
	AI1	TT	Temperature transmitter	Actual water temp
From operator	AI2	SLP	Set level potentiometer	Set desired water level
panel	AI3	STP	Set temp potentiometer	Set desired water temp
Not used	AI4–AI7			
Analog output module (8 channels: AO0–AO7)				
To process	AO0	VCV	Variable control valve	Actual water level
	AO1	VVD	Variable voltage drive	Actual water temp
To operator panel	AO2	LDM	Level display meter	Actual water level
	AO3	TDM	Temp display meter	Actual water temp
Not used	AO4–AO7			

Table 10.14 Allocation of Memory Locations for Special Requirements

Name	Function	Purpose
XXX		To store intermediate results

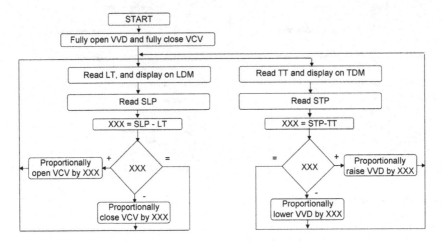

Figure 10.11 Loop control—flow chart.

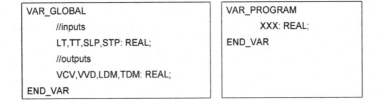

Figure 10.12 Allocation of variables.

Figures 10.12 and 10.13 illustrate the allocation of variables (global and program) and the coded program.

The program was coded and tested using CoDeSys software.

10.3.3 Two-Step Control with Dead-Band

Figure 10.14 illustrates a water heating process with its instrumentation subsystems (continuous input and discrete output devices).

Figure 10.15 illustrates the configuration of the controller for the automation of a water heating process (two-step control).

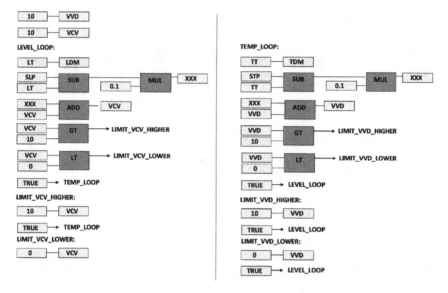

Figure 10.13 FBD program.

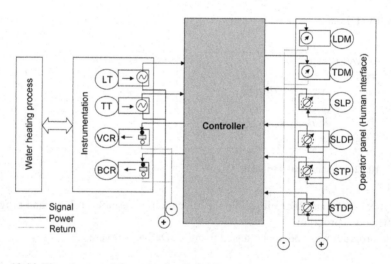

Figure 10.14 Water heater with automation (hybrid).

Table 10.15 illustrates the allocation of I/O channels to various inputs and outputs.

Table 10.16 illustrates the allocation of memory locations used for specific purposes.

Figure 10.16 illustrates the flow chart of the automation strategy.

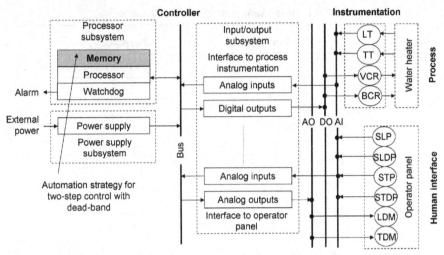

I/O modules can be shared by both process instrumentation and human interface subsystems.

Figure 10.15 Controller configuration.

Table 10.15 Allocation of I/O Channels to Process Parameters

From/To	I/O Channel		Instrumentation Device	Signal

Analog input module (8 channels: AI0–AI7)

From/To	I/O Channel		Instrumentation Device	Signal
From process	AI0	LT	Level transmitter	Actual water level
	AI1	TT	Temp transmitter	Actual water temp
From operator	AI2	SLP	Set level potentiometer	Desired water level
panel	AI3	SLDP	Set level dead-band potentiometer	Desired water level dead-band
	AI4	STP	Set temp potentiometer	Desired water temp
	AI5	STDP	Set temp dead-band potentiometer	Desired water temp dead-band
Not used	AI6-AI7			

Analog output module (4 channels: AO0–AO3)

From/To	I/O Channel		Instrumentation Device	Signal
To operator panel	AO0	LDM		Actual water level display
	AO1	TDM		Actual water temp display
Not used	AO2, AO3			

Digital output module (8 channels: DO0–DO7)

From/To	I/O Channel		Instrumentation Device	Signal
	DO0	VCR	Valve control relay	Open/close valve
	DO1	BCR	Breaker control relay	Close/open breaker
Not used	DO2–DO7			

Table 10.16 Allocation of Memory Locations

Name	Function	Purpose
XXX		To store intermediate results

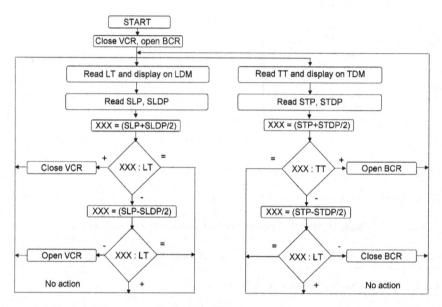

Figure 10.16 Two-step control with dead-band—flow chart.

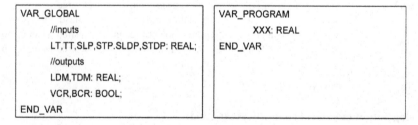

Figure 10.17 Allocation of variables.

Figures 10.17 and 10.18 illustrate the allocation of variables (global and program) and the coded program.

The program in Fig. 10.18 was coded and tested using CoDeSys software.

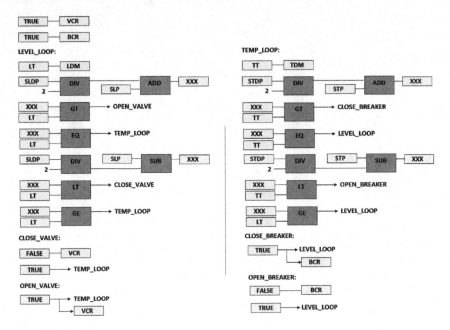

Figure 10.18 FBD program.

10.4 Summary

In this chapter, we studied the methods for programming the controller for customization for different applications. After touching upon machine-level and assembly-level programming, the advantages of higher-level language programming were discussed. It offers ease of programming, portability, availability of standardized languages, cross compiling in the host machine, and downloading into the target machine. The entire discussion was supported by basic programming symbols/instructions in ladder diagram and functional block diagram languages with simple application programming examples.

Finally, automation programs were presented for sequential control with interlocks (discrete water heating process), continuous control (continuous water heating process), and two-step control with dead-band (hybrid water heating process).

10.4 Summary

11 Advanced Human Interface

11.1 Introduction

In Chapter 5, we discussed the functions of the hardware-based human interface subsystem, or operator panel. Figure 11.1 illustrates the operator panel, which is integrated with the controller through input/output modules for human interface with the water heating process. The information is displayed on lamps and meters and the manual control is from switches and potentiometers, and they are all mounted on the panel at appropriate locations on the passive process diagram. The human interface functions are achieved through programming of the controller.

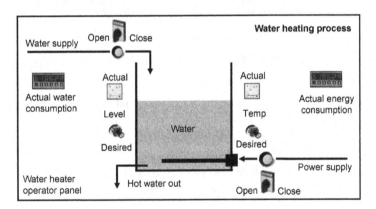

Figure 11.1 Operator panel.

The traditional operator panels, which are technically simple and less expensive, are still popular in many small or simple applications. However, they are not intelligent and have no provision for data exchange with the controller. They can only send and receive **electronic signals** to and from the controller through the I/O subsystem.

Other than sending and receiving electronic signals from the controller, there is also a need to input other information into the controller, such as date and time, so that the events in the process (alarms, measurements, control actions, etc.) are recorded and stored with time tags on a real-time basis.

In this chapter, the evolution, advantages, and disadvantages of advanced computer-based human interface subsystems are discussed. Advanced human interface subsystems are totally software-based and work with the latest in graphic user

Overview of Industrial Process Automation. DOI: 10.1016/B978-0-12-415779-8.00011-5

interface (GUI) techniques. They are called **operator stations**. The following sections discuss the transformation of the operator panel to the operator station.

Operator stations began as the application of monochrome alphanumeric display terminals and printers for operator interactions with the process, though these stations differed significantly from traditional operator panels. Apart from outputting the information on the display screen or printout, they provide keyboards for inputting the information into the terminals. Figure 11.2 illustrates typical display and printer terminals. These terminals support dot matrix for display and printing of characters.

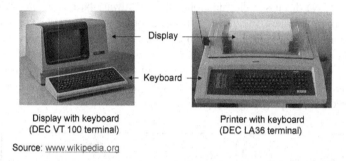

Display with keyboard
(DEC VT 100 terminal)

Printer with keyboard
(DEC LA36 terminal)

Source: www.wikipedia.org

Figure 11.2 Alphanumeric terminals.

The operator performs the following sequence of interactions or dialog with the controller through the alphanumeric display/printer terminal:

- Operator sends a query (inputting of a predefined text message) through keyboard of the terminal.
- Operator receives the echoes of the query by displaying or printing it on the terminal.
- Controller accepts the query, if it is valid.
- Controller processes the query.
- Controller responds (outputting of a predefined text message) through display/printing on the terminal.

The keyboard dialog and display/print messages are in easily understandable code (in the form of text) to address the relevant process parameters and the actions required. The interactive terminal, being only alphanumeric, does not support any graphics. These terminals are communicable, but they are also known as dumb terminals since they do not have any built-in intelligence, and they just function as an input/output device (inputs from the keyboard and outputs on the display screen/ printer paper) totally controlled by the program in the controller.

11.2 Intelligent Operator Panels

Since the cost and complexity of advanced human interface subsystems are high for simple applications, special interactive terminals are designed. Even though these are processor-based, these are replacements for alphanumeric display terminals and have intelligent operator panels with the following:

- Display functions have limited lines and limited characters per line, compared to 24/32 lines with 80 characters per line on a standard alpha-numeric display.
- Keyboard functions have limited keys for special operations, compared to 102 keys on a standard keyboard.

However, the terminal, by and large, maintains the same functionality as the traditional display terminal (monochrome display and keyboard for dialog). This provides a simple and cost-effective solution when needed. Since it is intelligent, it is possible to customize the terminal (programming display lines and key functions) for specific requirements. Further, the modern operator panels are equipped with serial and/or Ethernet interfaces for communicating with the controller, which allows them to be configured with specific functions for different keys or names for displays, etc. Figure 11.3 illustrates the typical structure of a modern operator panel.

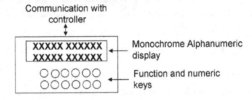

Figure 11.3 Modern operator panel—structure.

Figure 11.4 illustrates an industry example of an operator panel with an interface for communication with the controller. The interface is used not only for communication with the controller on a real-time basis but also for off-line configuration of the panel for customization of display and function keys. Its specifications are as follows:

- Monochrome LCD screen
- Two lines, each with 20 characters per line display
- Seven operator's keys (four of them customizable)
- Serial programming port for off-line configuration and online operation with a controller
- MS Windows-based software for configuration

Courtesy: www.schneider-electric.com

Figure 11.4 Operator panel—example.

11.3   Operator stations

With the advances in display technologies, the interactive terminals became more powerful and started supporting color graphics. Operator stations followed suit and adopted these developments in graphic user interface technology and became closer in appearance and operation to hardware-based mimic/control panels. Finally, low-resolution semi-graphic displays were replaced by high-resolution full graphic displays, as illustrated in Figure 11.5.

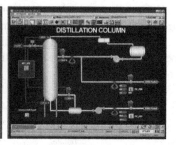

Courtesy: www.emerson.com

Figure 11.5 Full graphic displays.

With full graphic display terminals, the mode of operator interaction with the process shifted from keyboard/coded text message display to keyboard/mouse with cursor full graphic display. These changes are discussed in detail in the following sections.

Before we discuss an industry example of the operator station, let's revisit the automation of the water heating process and design an operator station for human interaction. In the water heating process with a two-step control, the following actions are taken on a continuous basis:

- Valve gets opened (to allow the cold water) whenever the water level in the tank goes below the desired value; otherwise, no action.
- Breaker gets opened (to disallow the power) whenever the temperature of the water in the tank goes above the desired value; otherwise, no action.
- The water and power consumption is computed.

Figure 11.6 illustrates the interconnections between the controller, instrumentation subsystem, and operator station.

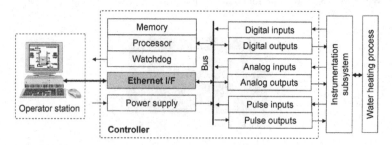

Figure 11.6 Operator station for water heating process.

The operator station is connected to the controller over an Ethernet communication interface and is programmed for the following basic functionalities:

- Real-time display of actual and desired values of level and temperature
- Real-time display of actual consumption of water and power
- Real-time display of state (opened and closed) of the valve and breaker
- Manual control for setting (increase or decrease) the desired values of level and temperature
- Manual control for changing the state (open or close) of the valve and breaker

Figure 11.7 illustrates the layout of the display on the screen and the control for operation of the previous functionalities.

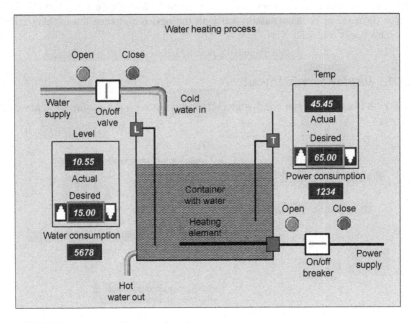

Figure 11.7 Display screen for water heating process.

The following are explanations of various passive and active components on the display screen:

- *Display screen layout*: Display screen background is a passive diagram with the layout/mimic of the water heating process.
- *Active display elements*: Bars within the block are provided for the display of the actual state of the valve and breaker (horizontal is open, vertical is closed). Numerical values are provided for the display of actual and desired values of level/temperature and water/power consumption.
- *Active control elements*: There are buttons on the screen to open or close the valve/breaker. Similarly, there are up and down arrows to increase or decrease the desired values.

For the following display operations, no procedure is needed, as the objects are continuously scanned and the displays are updated:

- Current values of level/temperature in numeric real form
- Desired values of level/temperature in numeric real form
- Current consumption of water/power in numeric integer form
- Current states of valve/breaker as position of the bar

For the following manual operations, a sequence of operations is needed, as they are discrete- or need-based:

- *Close/open valve/breaker*: With the cursor, select the specific button associated with the device and left-click for the action (open or close). The operator station ignores the invalid commands (issue of open command when the device is already open or issue of close command when the device is already closed).
- *Increase/decrease desired values of level/temperature*: With the cursor, select the specific arrow (increase or decrease) associated with the device and keep left-clicking until the desired value is reached.

11.3.1 Display Screen Layout

Figure 11.8 illustrates a typical display screen layout with the minimum features and facilities.

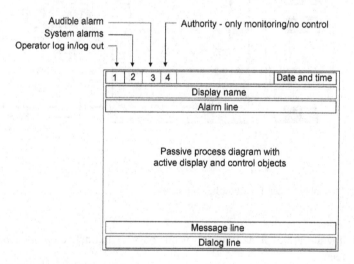

Figure 11.8 Display screen layout.

The explanations of various display fields are as follows:

- Operator log in/log out field displays whether the operator has logged into the operator station or logged out.
- System alarms field displays the presence of one or more automation system alarms.
- Audible alarm field displays whether this facility is enabled or disabled.
- Authority field displays the level of authority given to the operator—full control (monitoring as well as control of process parameters) or merely monitoring.

- Date and time field displays the current date and time.
- Display name field displays the name of the specific display.
- Alarm line field displays the latest alarm.
- Passive process diagram with active display and control objects field displays the mimic of the process or process diagram, with active display and control objects in the appropriate places for human interaction.
- Message line field displays the system messages.
- Dialog line field displays the echo of the operator dialog through the keyboard

The display screen layout varies from system to system and vendor to vendor. Modern systems provide a lot more facilities than described here.

11.3.2   Interaction with the Process

There are many ways to interact with the process through operator stations. Some typical cases are discussed in this section.

11.3.2.1   Direct Interaction

Direct interaction with the process using the operator station is done for display of the values/states of the process parameters or for manual control of the values/states of the process parameters. The sequence of operations involved is as follows:

- *Display of process values/states*: No interaction with the process is required, as all the values/states of interested parameters are continuously displayed on the screen.
- *Control of process values/states*: The following sequence is used to perform the action:
  - Identify the control symbol associated with the object by positioning the cursor on its symbol on the display screen. The control symbols are open/close and increase/decrease.
  - Execute the operation by selecting the object with a left-click.

Figure 11.9 illustrates the application of this approach for the water heating process.

11.3.2.2   Navigated Interaction

The simple or straightforward approach is not always possible when more information needs to be displayed and/or more process parameters need to be controlled with many options. This leads to overcrowding on the display area. To overcome this, a navigated interaction with the process can be employed, which is how normal computers work. To explain this scheme, let's remove simple approach features for control and set-point facility and replace them with two level navigation facilities, as illustrated in Figure 11.10.

The sequence of operations adopted for controlling the breaker or the valve is as follows:

- Identify the object to be controlled (breaker or valve) by placing the cursor on the object.
- Left-click on the identified object for its selection.
- Right-click on the selected object for getting the main options (open or close).
- Right-click on the selected main option (open or close) for getting secondary options (execute or cancel).

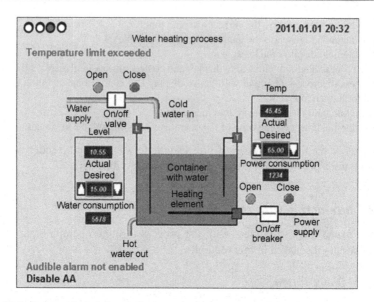

Figure 11.9 Operator interaction—direct.

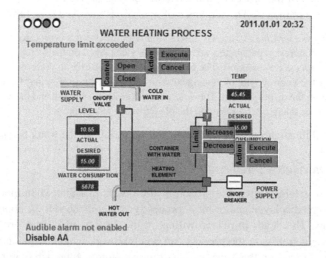

Figure 11.10 Operator interaction—navigated.

- Place the cursor on the selected secondary option (execute or cancel).
- Left-click on the selected option (execute or cancel) to take the final action (execution of the command or to cancel the command).

Similarly, the sequence of operations adopted for changing the limit of the temperature or the level is as follows:

- Identify the object to be controlled (desired level or desired temperature) by placing the cursor on the object.
- Left-click on the identified object for its selection.
- Right-click on the selected object for getting the main options (increase or decrease).
- Right-click on the selected option (increase or decrease) for getting secondary options (execute or cancel).
- Place the cursor on the selected secondary option (execute or cancel).
- Left-click on the selected option (execute or cancel) to take the final action (execution of the command or to cancel the command).

In both of the previous cases, if the operator makes a mistake (e.g., trying to close the breaker or valve when they are already in the closed state), the system indicates the unavailability of this option (deactivated or disabled with a faded display) and does not accept the command in the main level itself.

The sequence we discussed for operator interactions with the process is called navigation. In the previous example, only two levels are discussed. In practice, there may be many levels in the tree structure. The structure and the number of levels depend on the nature of the object and its attributes.

Figure 11.11 illustrates a navigation menu with three levels (main and two secondary levels).

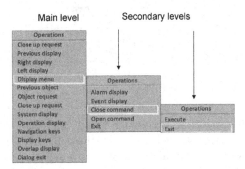

Figure 11.11 Multi-level navigation.

Multi-level navigated interaction with the process is summarized below and is applicable for both display and control for manual interaction:

- Identify the desired object by positioning the cursor on its symbol in the display screen.
- Select the identified object with a left-click to get the main menu.
- Select the option in the main menu with a right-click to go to the secondary menu.
- Select the option in the secondary menu with a right-click to go to the next level of the secondary menu.
- Continue until the last level of the secondary level menu.
- Select the required operation within the menu in the last level and left-click for its execution.

When this sequence of operations is complete, the selected operation gets executed.

Figure 11.12 illustrates an example of a display of an electrical substation with the measurements and status of the process objects displayed. In this case, the simple approach is used for viewing the values and states of various substation parameters. The isolators and circuit breakers (represented by I1, I2, and so on, and C1, C2, and so on) indicate their state using symbols that are either blank (device is open) or filled with color (device is closed). Similarly, values of various analog parameters (current, voltage, MW, MVAR, tap position of transformer) are displayed next to the device.

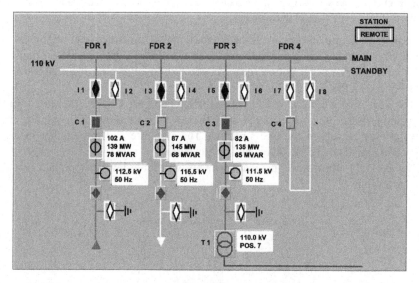

Figure 11.12 Display for electrical substation.

Figure 11.13 illustrates the navigation-based approach for process control. Here, the operator controls the devices (open/close of isolators and circuit breakers, increase/decrease of transformer taps) through a dialog window associated with the device. By positioning the cursor on the object and right-clicking, the device is selected and a dialog box appears. For issuing a command, the option within the dialog box is selected and the command is executed.

Transformer tap change

Open/close circuit breaker

Figure 11.13 Command execution.

11.3.2.3 Other Features

Modern operator stations provide the facility to call for the information related to the desired object, both for viewing as well as for control. An example featuring a faceplate is illustrated in Figure 11.14. The faceplate displays the status and data of the selected object, and it also provides for the entering of data, changing of modes, etc.

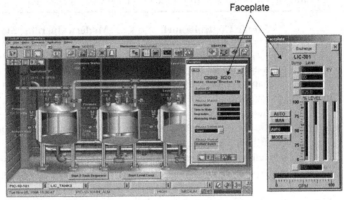

Courtesy: www.emerson.com

Figure 11.14 Faceplate.

In addition to the facilities discussed so far, the operator station also provides following additional displays, as illustrated in Figure 11.15, which are not possible in traditional mimic-based operator panels:

- Alarm/event message display
- Parameter trend display
- Multiple windows display

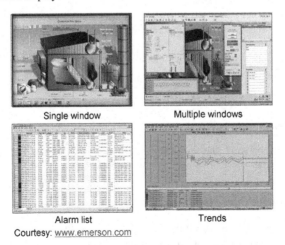

Courtesy: www.emerson.com

Figure 11.15 Additional displays.

11.3.3 Comparison with Operator Panel

Table 11.1 compares the functions of the operator station with those of the traditional mimic control panel.

Table 11.1 Comparison of Mimic Panel and Operator Station

Functions	Mimic Control Panel	Modern Operator Station
Appearance	Passive process display diagram with the active panel components at relevant locations on a metal/fiber or mosaic panel	Passive process diagram with symbols for active objects at relevant locations on the display screen
Display of information	On display components, such as lamps, meters, recorders, counters, dedicated to the process parameters	Near or on active symbols, such as discrete and continuous objects, dedicated to the process parameters on the screen
Manual operation	Physical operation of control components, such as push buttons, potentiometers, mounted on the panel	Use of menus and submenus for navigation for issue of commands, and for displays

11.3.3.1 Advantages and Disadvantages of Operator Stations

The advantages are as follows:

- Consumes less space and power.
- No I/O subsystem and associated wiring is necessary for connecting the controller with the human interface subsystem, as the operator station is interfaced to the controller over an Ethernet communication interface.
- The human interface subsystem is flexible, allowing expansion and/or modification.
- No operator movement is necessary from the control desk for control execution on the operator panel.
- Can also display additional information, such as alarm/events, multiple windows, and trends.

The disadvantage is as follows:

- The only disadvantage is that the operator station cannot clearly display the panoramic view of the large processes in view of its limited display area.

11.3.4 Enhanced Operator Stations

Enhanced operator stations are employed to overcome the unavailability of the panoramic view of the process, as discussed in the following sections.

11.3.4.1   Multiple Monitors

In the arrangement shown in Figure 11.16, the operator station can have more than one monitor for interaction with the process, or a multi-monitor-based interaction. For instance, one monitor may be for control and interaction with the process while the other monitors are for the display of information that is required most of the time. This arrangement, equipped with a special multi-monitor graphic controller, uses a single keyboard and single mouse with a flying cursor that can jump from one screen to the other as desired.

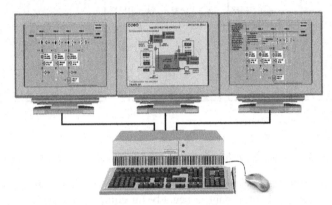

Figure 11.16 Multi-monitors with a common keyboard and mouse.

Figure 11.17 illustrates an industry example of an operator station with dual and quad monitors with a single keyboard and mouse.

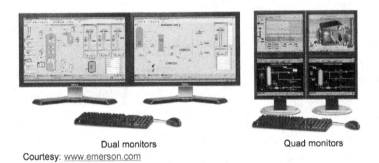

Dual monitors   Quad monitors
Courtesy: www.emerson.com

Figure 11.17 Dual and quad monitors with a common keyboard/mouse.

11.3.4.2   Large Screen Displays

Multiple monitors are arranged in a matrix to get a large screen display of the process (distributed on all the monitors). Figure 11.18 illustrates a display over nine monitors (3×3 matrix). Normally, such displays are provided in the background in the control center and are not used for routine interaction with the process.

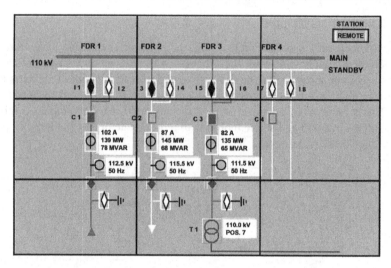

Figure 11.18 Matrix of monitors.

11.3.4.3 Displays with Embedded Video

Current automation systems support viewing live video streams from remote cameras distributed throughout a plant, especially for equipment surveillance for safety, security, emissions, etc., and for tracking assets, people, etc. The video streams are integrated with other process displays on operator stations. This is possible because of the interfacing and remote controlling of networked cameras installed throughout plant. An industry example of display with embedded video, via remotely controlled cameras, is illustrated in Figure 11.19.

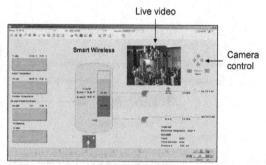

Tilt-zoom remote camera
Courtesy: www.emerson.com

Display with integrated live video

Figure 11.19 Display with embedded video.

11.3.4.4 Combined Mimic Panel and Operator Station

As illustrated in Figure 11.20, the mimic panel with only a few important active display elements is used for the panoramic view while the normal operator stations are used for regular interaction with the process.

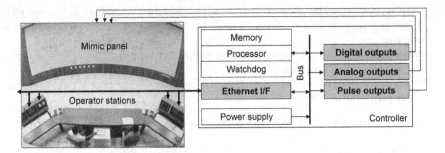

Figure 11.20 Combination of mimic panel and operator station.

11.3.5   Variants of Operator Stations

Process operators use operator stations only for monitoring and control of the process. **Maintenance stations** are provided for maintenance of the automation system and **engineering stations** are for making modifications or reconfigurations in the automation system. Except for their display and interaction facilities, these stations are functionally similar to the operator station.

11.4   Logging stations

Another very important component of the modern human interface subsystem is a logging station. A logging station is basically a printer (generally combined with graphic capabilities to support the printing of text, displays, graphs, etc.). The logging stations are generally used for data logging, as detailed in Section 11.4.1.

11.4.1   Data Logging

Data logging means dumping process data in a required format. The dump can be on a printer or on any other media, such as a hard disk. The following are some examples of different types of logging.

Alarm/event logging: Here, the automation system is programmed to print out a line of information with the details whenever an alarm or an event takes place in the system. Operator dialog with the system is included as an event. Each line typically has several information fields, such as the following:

- Date
- Time (up to one millisecond resolution)
- Type of event
- Category of the event
- Area of the event
- Name of the parameter
- State of the event

Trend logging: Here, the automation system is programmed to print out continuous plots of process parameters over short time intervals for close observation of the selected process parameters.

Report logging: This includes hourly, daily, and weekly reports on the process variables (including the computed variables).

On demand logging: The operator can request a log either on display or in print of the current display, trend plot, etc.

Typical log lines are illustrated in Figure 11.21.

Date	Time	Type	Category	Area	Parameter	State
2011/01/28	20:12:19:300	Event	Process	Boiler	Temp 15	High
2011/01/28	20:12:20:300	Alarm	Process	Turbine	Vibration	High
2011/01/28	20:12:22:300	Command	User	Substation	Circuit breaker	Closed

Figure 11.21 Alarm/event log lines.

For logging, the printers are either connected to an individual operator station (dedicated) or placed on the local area network (LAN) for sharing by all the operator stations. This is illustrated in Figure 11.22.

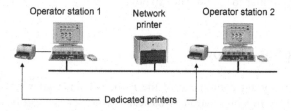

Figure 11.22 Logging stations with dedicated and shared printers.

11.5   Control Desk

The control desk is a physical arrangement of operator stations and furniture, as illustrated in Figure 11.23. This arrangement takes care of all **ergonomic** requirements so that the operators are able to interact with the process without fatigue.

Courtesy: www.emerson.com

Figure 11.23 Control desk.

11.6 Summary

In this chapter, we discussed the operator station, a modern human interface sub-system based on graphic user interface (GUI) technology. Even though the operator station does not offer the panoramic view of the large process, which may still be required in some cases, they do provide many other benefits to compensate for this particular shortcoming. Operator stations may also have a large screen display, but the display functions remain the same.

12 Types of Automation Systems

12.1 Introduction

In Chapter 1, we discussed the following two types of physical processes:

- Localized processes (present over a small area)
- Distributed processes (present over a large/geographical area)

Figure 12.1 illustrates a typical electricity distribution process (localized as well as distributed) in a municipal area.

The overall structure of the systems required for the automation of the two different types of processes differ significantly in nature. In the following sections, the automation systems specially devised for these process types are discussed.

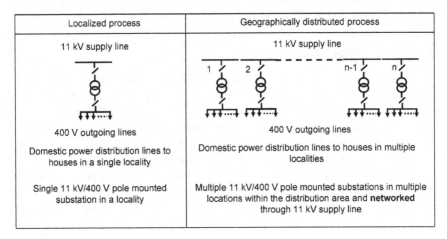

Figure 12.1 Electricity distribution process.

12.2 Localized Process

The local process are present in a relatively small physical area, and the control center is physically close to the process. Figure 12.2 illustrates the structure (logical and physical) of the automation system for the localized process. Here, we see that the controller is locally connected to the operator station over the local communication line through the communication interface. Normally, an Ethernet interface is

Overview of Industrial Process Automation. DOI: 10.1016/B978-0-12-415779-8.00012-7

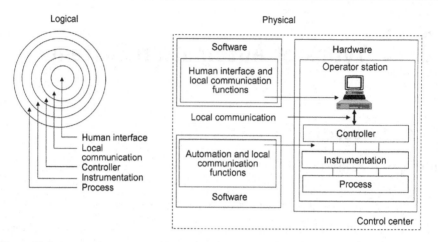

Figure 12.2 Automation system structure—localized process.

employed for communication between the controller and the operator station. The automation systems for the localized process can be either centralized or decentralized or distributed, as explained in the following sections.

12.2.1 Centralized Control System

A centralized control system (CCS) always employs a single communicable controller that takes the full automation load of the entire process. However, the system can have either a single operator station or multiple operator stations. In the case of CCS with multiple operator stations, multiple operators simultaneously share interaction with the process. This sharing improves operator efficiency, especially when the process is large. The sharing is generally either on a functional basis or on an area basis.

Let's use the example of a power plant automation system. A power plant typically has three functional subprocesses—boiler, generation, and auxiliary—that produce electricity from coal and water, as illustrated in Figure 12.3. With a single large

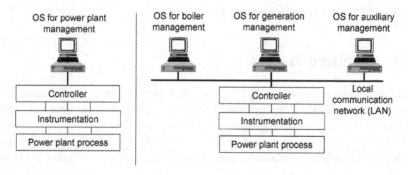

Figure 12.3 CCS with single and multiple operator stations.

controller, the entire plant automation, with single or multiple operator stations, can be used for operator interaction.

Important advantages and disadvantages of CCS are as follows.

Advantages:
- Ideal for small and less spread out processes because of minimum field cabling between the instrumentation/process points and the controller
- Technically simple
- Less expensive

Disadvantages:
- Not economical for large and widely spread out processes because of extensive field cabling between the instrumentation/process points and the controller
- Failure of the controller results in making the complete plant automation facility unavailable
- Handling, troubleshooting, maintenance, etc., of a big controller are very inconvenient

12.2.2 Decentralized/Distributed Control System

To overcome the problems associated with CCS for large or widely distributed processes, a network of many distributed communicable controllers is formed. These controllers share the full automation load of the entire process, as shown in Figure 12.4. Here, both the controllers and operator stations share the load on a functional or area basis. As all the controllers and operator stations are on the same network, the information in one controller can be shared by other controllers and/or operator stations. This arrangement is called a decentralized or distributed control system (DCS).

A DCS is a local area network of controllers and operator stations.

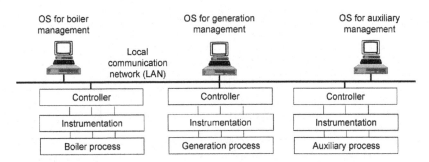

Figure 12.4 DCS with multiple controllers and operator stations.

The concept of a DCS is best explained using the example of the automation of the power plant process with its three functional subprocesses—boiler, generation, and auxiliary. The three smaller communicable controllers for the three subprocesses control the complete process jointly. Important advantages and disadvantages of DCS are as follows.

Advantages:
- Ideal for large and widely spread out processes (minimum cabling between the process/ instrumentation points to the distributed controllers, as the controllers can be placed closer to the subprocesses)
- Higher overall availability, as single controller failure does not lead to a total automation system failure (affects only a function or a part of the plant)
- Each controller being smaller in size, handling, troubleshooting, maintenance, etc. are relatively convenient

Disadvantages:
- Expensive
- Technically more complicated

12.3   Distributed Process

The distributed process, a group of interconnected localized subprocesses, is spread out over a relatively large physical (or even geographical) area, and the control center is physically away from the subprocesses. Figure 12.5 illustrates the structure (logical and physical) of the automation system for one localized subprocess in a distributed process. The controller is remotely connected to the operator station over the remote communication line through the communication interface. Normally, a serial communication interface is employed for communication between the controller and the operator station. The automation system for the distributed processes can be a simple remote control system (RCS) or a large network control system (NCS), as explained in the following sections.

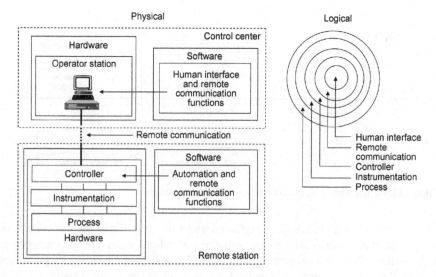

Figure 12.5 Automation system structure—distributed process.

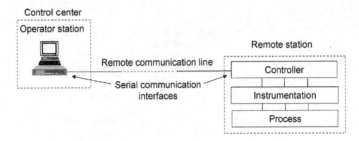

Figure 12.6 Remote control system.

12.3.1   Remote Control System

In a remote control system, as shown in Figure 12.6, the operator station located in the control center monitors and controls the remotely located process over a communication line. The controller, connected to the instrumentation and the process, treats the process as localized. It performs its automation functions locally without the assistance of an operator station, and it communicates with the operator station in the control center. The role of the operator station is only to supervise and monitor the process and issue manual control commands, if required. The RCS, generally, employs serial communication interfaces at both ends. The system in the control center is often called a master station.

12.3.2   Network Control System

A network control system is an extension of RCS, and it simultaneously monitors and controls many geographically distributed localized processes from a central place. Multiple communicable controllers, distributed geographically and interfaced to their own instrumentation and processes, share the common long-distance communication network for communication with the control center (operator station). Each controller manages its own local automation independently, while the operator station in the control center supervises the overall automation of the entire distributed process. Further, each controller can communicate with the control center only and not with other controllers in the network. In other words, the control center is always the master of the communication network, and any communication among the controllers is always through the control center. Figure 12.7 illustrates the structure of an NCS.

Further, the network in an NCS refers to the network of interconnected distributed localized processes (not the communication network). For example, the electrical transmission network is a distributed process where the localized processes are the substations, and networking is through the transmission lines.

12.3.3   Front-End Processor

In Figure 12.8, the operator station in the RCS/NCS performs two functions: human interface and remote communication. Normally, the communication links over serial

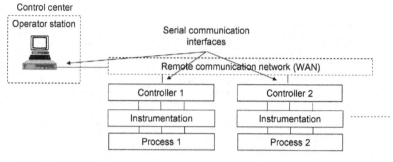

Figure 12.7 Network control system.

interface are slow and less reliable due to remote connection. Often, the arrangement calls for more repeat data transmission compared to its local communication equivalents. Ethernet communication links, though faster and more reliable, are not used for long-distance remote communication.

In the NCS with many remote stations connected, the remote communication demands extensive processing time from the operator station, reducing its performance resulting in slow response of operator interactions. To overcome this, the routine communication load is shifted to an independent platform called a front-end processor (FEP), as illustrated in Figure 12.8. The advantage of FEP is that it

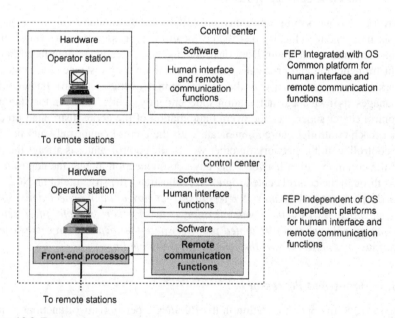

Figure 12.8 Front-end processor.

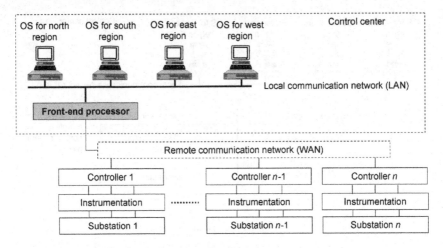

Figure 12.9 Electricity transmission network automation.

improves the performance of operator stations, while the disadvantages are that they are expensive and technically complex.

Figure 12.9 illustrates the implementation of an NCS for the automation of an electricity transmission network in a region employing independent FEP. As mentioned earlier, this is a network of geographically distributed substations through which the electrical transmission lines pass.

A general NCS has two communication networks:

* A local area network (LAN) of operator stations and FEP within the control center
* A wide area network (WAN) of controllers and FEP outside the control center

FEP not only takes care of the routine communication load between the control center and the distributed controllers, but also provides a buffer between high-speed LAN and low-speed WAN communication. Further, depending on the processing power, FEP can also be used to preprocess the incoming data from controllers before sending it to the operator station and to prepare the data received from the operator station before dispatching to the distributed controllers.

FEP can be realized by using either the controller platform or a general-purpose computer platform.

12.3.3.1 Controller-Based FEP

Figure 12.10 illustrates a process controller-based FEP in which the controller is configured with the following:

* Ethernet communication interface for local communication between FEP and operator stations
* Serial communication interface for remote communication between FEP and geographically distributed controllers
* Memory resident program for local as well as remote communication handling

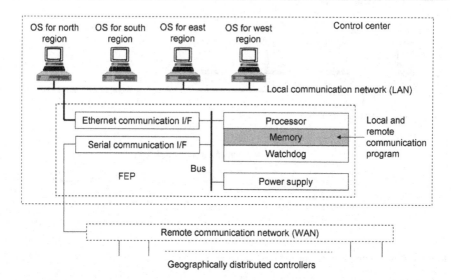

Figure 12.10 Controller-based FEP.

The advantage of the controller-based FEP is that the system is based on a rugged hardware platform and is maintenance-free. The disadvantage is its limited memory and processing capabilities, which are specially designed for automation functions, have limited facilities for supporting communication functions.

12.3.3.2 Computer-Based FEP

Figure 12.11 illustrates a general-purpose computer-based FEP configured with an Ethernet communication interface for local communication with operator stations, a serial communication interface for remote communication with geographically

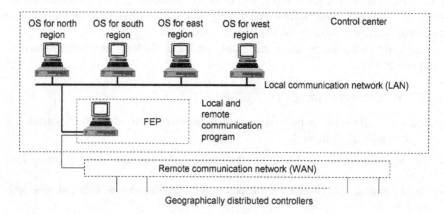

Figure 12.11 Computer-based FEP.

distributed controllers, and a memory resident program for local as well as remote communication in line with the controller-based FEP.

The advantage of a general-purpose computer-based FEP is that it provides an excellent platform for supporting the communication function with no limitations on memory, computing power, etc. The main disadvantage is that it is not maintenance-free and has higher overhead, which are not desirable in automation systems.

It is also possible to employ NCS for a localized process in place of DCS, provided the functional, cost, and performance requirements are satisfied.

12.4   Supervisory Control and Data Acquisition

Supervisory control and data acquisition (SCADA) is an application of automation technology where there is a need for an independent arrangement to coordinate distributed controllers in automation systems (both in DCS and NCS), as explained further in the following sections.

12.4.1   Background

In the early days, before the arrival of communicable controllers, the following was scenario:

- Only the hardwired/stand-alone controllers were employed for automation of each subprocess in a complex process.
- Each hardwired/stand-alone controller operated, generally, on a single input/single output (SISO) basis.
- Hardwired/stand-alone controllers, which were not intelligent and not communicable, did not support any interaction among themselves for coordination.
- The computers did not enjoy the confidence of the operating personnel (not meeting the acceptable level of reliability).
- Computers, as an intermediate step, were used as supervisory tools only to monitor, coordinate, and optimize the functions of stand-alone/hardwired controllers.
- The failure of the computer did not affect the process control, as the controllers retained their set-points and functioned independently of the computer.

The above is explained below for the management of a group of generators to meet specific requirements.

12.4.2   Case Study

In a typical power grid, all the synchronous generators connected to the grid operate at the common grid frequency. They remain synchronized to the grid and share the total grid load proportional to their individual capacities. Each generator takes its share of the load while complying with the common grid frequency. Whenever the consumption is more than the generation, the grid frequency falls and causes the generators to take the additional load proportionally at a reduced grid frequency.

As shown in Figure 12.12, all the generators are equipped with speed controllers that are hardwired, noncommunicable, stand-alone, and autonomous. Each controller,

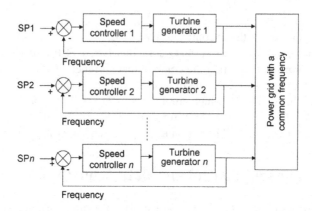

Figure 12.12 Independent operation of generators.

with its own set-point, regulates its turbo-generator's speed/frequency. There is no coordination among the controllers except the indirect influence by the grid frequency.

Often, generators are forced to take additional loads beyond their share of the total. During the outage of a generator, for example, some generators are forced to take on an extra load by operating at higher set-points. As the generators have to remain synchronized, their speed/frequency cannot change. Hence, their higher frequency falls to the grid frequency by generating more power. This way, the generators take higher shares of the load but still remain synchronized to the common grid frequency. SCADA principles come in very handy for the management of this situation, as illustrated in Figure 12.13. In this configuration, the supervisory computer

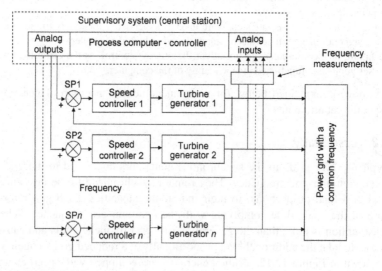

Figure 12.13 SCADA for power plant operation.

computes the best set-points for each generator to share the load and still remain synchronized with the grid. Failure of the supervisory computer does not affect individual speed controllers, as the set-points remain unchanged. This way, the supervisory system coordinates the operation of all the speed controllers.

12.4.3 Similarities with DCS and NCS

Even today, a network control system, with its distributed architecture, is still referred to as a SCADA system because of their similarities. While the distributed controllers perform the local automation functions independently, the control center just supervises the functioning of the distributed process and performs human interface functions. The failure of the control center and/or the remote communication network does not affect the automation functions of individual distributed remote stations.

Similarly, the hierarchical system structure with supervisory functions in a distributed control system also resembles a SCADA system. *SCADA is not a technology but an application.* Today, the term SCADA is treated as generic since all of its features and functionalities are available in both hierarchical DCS and NCS, as explained in the following sections.

12.4.3.1   SCADA in DCS

As shown in Figure 12.14, the supervisory computer, with common functions, coordinates the functions of the individual controllers. At the power plant level, it coordinates the boiler, generation, and auxiliary subprocesses. This is an example of SCADA in DCS.

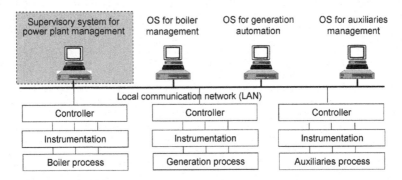

Figure 12.14 SCADA in DCS.

12.4.3.2   SCADA in NCS

As shown in Figure 12.15, the supervisory computer, with common functions, coordinates the functions of the distributed controllers at control center level, an example of SCADA in NCS.

In this example, SCADA only coordinates the functioning of controllers.

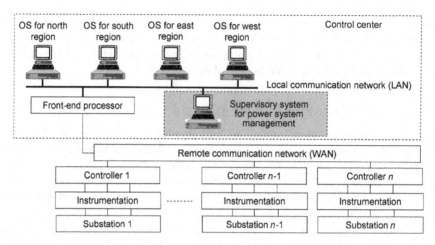

Figure 12.15 SCADA in NCS.

12.5 Summary

In this chapter, we discussed different types of automation systems for both local-ized and distributed processes. Centralized systems and decentralized systems were described for localized processes, remote control systems, and network control sys-tems for geographically distributed processes. Finally, SCADA functionality was discussed in both DCS and NCS.

13 Special-Purpose Controllers

13.1 Introduction

We discussed the types of automation systems in Chapter 12. In Chapter 7, we covered the general-purpose controller, its application, and customization for the automation of various processes (discrete, continuous, and combination). In this chapter, we further discuss the standardization of controllers to address the specific needs of localized and distributed processes.

The localized and distributed processes need functionally different and diversified types of controllers (different hardware and/or software configurations of general-purpose controllers), even though they appear physically similar. Different processes require different types of automation strategies. Generally, discrete processes require sequential control with interlocks, while continuous processes require closed loop control. A hybrid process requires both.

As explained in the following sections, all special-purpose controllers are communicable and can co-exist with others over a compatible communication network.

13.2 Controller for Localized Processes

The automation structure of the localized control system is illustrated in Figure 13.1.

The emphasis here is on local automation tasks. Following are the variants of general-purpose controllers for the localized processes:

- Programmable logic controller (PLC)
- Loop controller
- Programmable controller (Controller)

The following sections explain in detail the differences and application areas of special-purpose controllers.

13.2.1 Programmable Logic Controller

Programmable logic controllers were developed as stand-alone replacements for traditional relay logic (or its solid state equivalent, as explained in Appendix A) for discrete process automation. In principle, this is a controller configured with only digital inputs and outputs capable of executing only logic functions to support sequential control with interlocks. The main performance requirement here is

Overview of Industrial Process Automation. DOI: 10.1016/B978-0-12-415779-8.00013-9

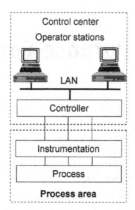

Figure 13.1 Automation system for localized process.

the **speed of execution** of the automation strategy (control logic). The hardware and software are specifically designed to address this requirement.

The current PLC is designed with high-performance processors, which also support analog data handling and exploit the available/built-in computing power of the processor. All the general features of the PLC, including its name, have remained unchanged over time, and the term *PLC* has become somewhat generic. Today's PLC supports limited continuous process automation while according highest priority and speed to its basic functions (logic processing for discrete process automation). Also, the modern PLC is communicable and can become part of a network of automation equipment.

Figure 13.2 illustrates an industry example of a small PLC with integrated inputs and outputs.

Courtesy: www.schneider-electric.com

Figure 13.2 Programmable logic controller.

Figure 13.3 illustrates a simple PLC-based water level indicator and controller for application in domestic buildings and apartments.

Depending on the water levels in the sump and overhead tank, continuous supply of water is ensured to the residents automatically, subject to the safety conditions. The functions of the equipment are detailed as follows:

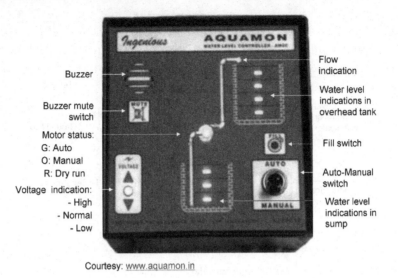

Courtesy: www.aquamon.in

Figure 13.3 PLC-based water level indicator and controller.

- *Display on operator panel*: Flow indication, level indications in sump and overhead tank, supply voltage status indications (high, low, normal), and motor status indication (manual, auto, dry rum)
- *Manual control on the operator panel*: Auto/manual selection, buzzer enable/disable (mute) selection, and fill operation
- *Interlock conditions*: Motor automatically starts when the water level in the overhead tank goes below the lowest level, provided the following interlock conditions are satisfied:
    - The auto/manual switch is on auto mode.
    - Supply voltage to the motor is within the limits, or normal.
    - Water level in the sump is above the lowest level.

The details of the additional interlock conditions are as follows:

- *Sump interlock*: To avoid the possibility of the motor switching on/off continuously (hunting) until the overhead tank is above low level, the motor is made to start only once the water reaches the second level in the sump.
- *Overhead tank interlock*: Motor can also start if sump reaches full level and tank level is below high level (100%). Motor does not start again for this condition until the overhead tank falls below 75%.

- *Actions*: Motor automatically stops whenever any of the following conditions are detected:
    - Overhead tank full
    - Sump empty
    - Motor dry run
    - Abnormal voltage
    - Both overhead tank and sump empty

This is an example of sequential control with interlocks implemented with PLC. The implementation of the application of this product is illustrated in Figure 13.4.

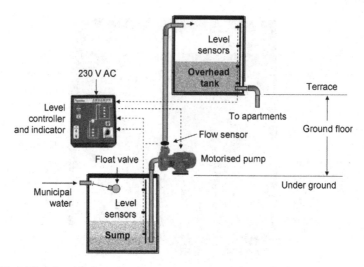

Figure 13.4 Water level indicator and controller.

13.2.2 Loop Controller

The processor-based loop controller is designed as a replacement to the earlier stand-alone solid state analog loop controllers. This is also a special-purpose controller to support continuous process automation (closed loop control). Unlike the traditional PLC, which can handle only the logic functions, the loop controller is designed to handle arithmetic and mathematical functions in the context of continuous control. An important performance requirement of the loop controller is its ability to handle the analog data with **accuracy**. As the analog process parameters vary relatively slowly in real-time, speed of execution is not a very important criterion. Hence, the hardware and software are specifically designed to address the accuracy requirement.

Being processor-based, modern loop controllers also provide a few (very limited) digital I/Os functionally related to the basic analog I/O. Further, the loop controllers are also used as **backup systems** for critical loops in continuous process automation in case of the failure of the DCS (main/integrated control system). Modern loop controllers are communicable, have their own human interfaces, can be a part of a larger network of automation equipment, and can support additional control functions, as discussed in Appendix F.

Figure 13.5 illustrates an industry example of a single loop controller with an integrated operator interface.

Figure 13.6 illustrates a single loop controller's simple application of regulating the liquid flow in accordance with the set-point value. The controller receives analog inputs (continuous actual flow from the flow transmitter and desired flow or set-point from the operator panel), and it generates a continuous control signal to operate the motorized valve to regulate the liquid flow (following the reference value or reducing the deviation).

Courtesy: www.yokogawa.com

Figure 13.5 Single loop controller.

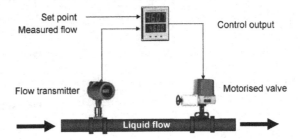

Figure 13.6 Liquid flow regulation.

Industry also provides for dual-loop and multi-loop controllers for specific applications, as discussed in Appendix A.

13.2.3 Programmable Controller

The programmable controller, usually just referred to as **controller**, is technically a functional combination of a PLC and loop controller that implements both discrete sequential control with interlocks and continuous control, which facilitates both discrete and continuous process automation. In continuous processes, even though the main requirement is continuous process automation, there are also discrete subprocesses that need automation. In view of this, the controller supports both digital and analog I/O. However, this is a general-purpose controller with more emphasis on continuous process automation and higher accuracy in handling analog information.

Figure 13.7 illustrates an industry example of a controller.

Figure 13.8 illustrates an application of the controllers in DCS.

Figure 13.9 illustrates the application of DCS for power plant automation. Here, all the subprocesses in the power plant are managed by a set of controllers and operator stations on the network (LAN).

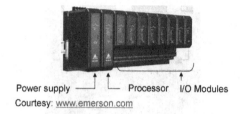

Power supply ──┐ └── Processor   I/O Modules
Courtesy: www.emerson.com

Figure 13.7 Controller.

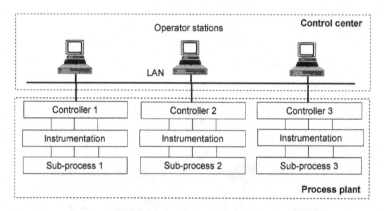

Figure 13.8 Controllers in DCS.

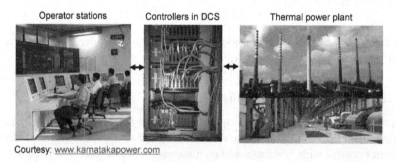

Courtesy: www.karnatakapower.com

Figure 13.9 Application of DCS in power plant automation.

13.3   Controller for Distributed Processes

The automation structure of remote controlled processes is illustrated in Figure 13.10. The industry name for this special-purpose controller is **remote terminal unit (RTU)**.

The emphasis here is on data communication/exchange between RTU and the remote control center. The RTU, traditionally, is designed to perform the following functions.

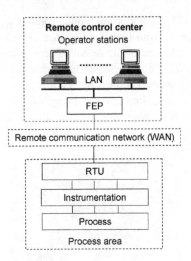

Figure 13.10 Remote control system.

From RTU to remote control center:

- Acquisition of values of continuous parameters locally from the process and transmission of these over the communication network for further processing. This remote data acquisition operation is known as **tele-metering** or **tele-measurement** (acquisition of analog process parameters).
- Acquisition of states of the discrete parameters locally from the process and transmission of these over the communication network for further processing. This remote data acquisition operation is known as **tele-indication** or **tele-signaling** (acquisition of discrete process parameters).

From remote control center to RTU:

- Sending control commands for discrete parameters of the process over the communication network and passing these on locally to the process. This remote control operation is known as **tele-control** or **tele-command** (execution of discrete process control).
- Sending set-points of continuous parameters of the process over the communication network and passing these on locally to the loop control function within the controller. This is also known as **tele-control** or **tele-command** (execution of analog process control).

Here, the term *tele* means performing an operation from a distance. This can be seen in the words *telephone, television, teleprinter,* etc.

13.3.1 Remote Terminal Unit

Early RTUs mainly supported data acquisition from the process and its transmission to remote control centers, as well as receipt of data from remote control centers and transfer to the process. Earlier versions did not support any built-in intelligence or capabilities, such as for local execution of the automation functions.

As in the case of preset PLCs supporting limited continuous process automation functions, present RTUs, designed with high-performance processors, support limited local automation functions to exploit the available built-in computing power. However, all the general features of RTU, including its name, have remained unchanged over time. In other words, today's RTUs support limited local automation functions while according highest priority to their basic functions (data exchange with the remote control center).

In practice, RTUs do not generally use the local automation functions even though these can be implemented. Figure 13.11 illustrates industry example of RTU.

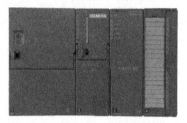

Courtesy: www.siemens.com

Figure 13.11 Remote terminal unit.

All RTU functions were earlier performed by set of independent tele-signaling, tele-metering, and tele-control equipment. But in the current approach, these functions are carried out by RTU. A comparison between the two approaches is illustrated in Figure 13.12 for one remote station. The number of unidirectional communication links required in the earlier approach increases by multiples of 3 with the addition of each remote station. However, in an RTU-based approach, it is possible to share the common bidirectional communication link among several RTUs.

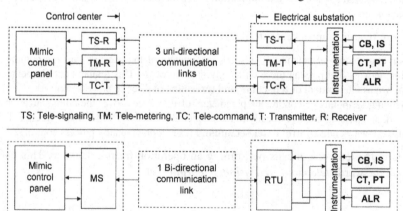

Figure 13.12 Approaches for remote monitoring and control.

Figure 13.13 illustrates a typical application of RTU for remote monitoring and control of an electrical substation.

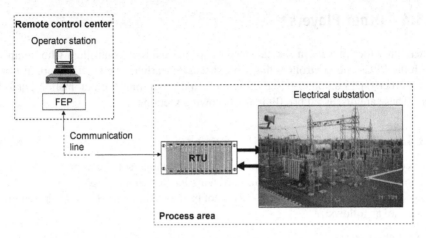

Figure 13.13 Remote management of electrical substation.

Tele-metering, tele-signaling, and tele-controlling concepts originated in electrical transmission and distribution systems because of the fact that these are large, geographically distributed processes. There was a necessity for managing the process from a central location. To start with, power lines were employed for communication in the form of power line carrier communication (PLCC). The communication medium for data was the same as that for speech and protection in the electrical transmission and distribution companies. The application later migrated to an RTU-based approach. Presently, the NCS approach is also being applied to other geographically distributed processes, such as oil/gas transmission systems and water transmission and distribution, as illustrated in Figure 13.14.

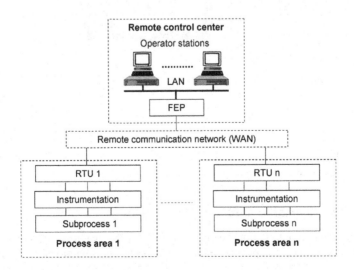

Figure 13.14 Application of RTU in NCS.

13.4 Other Players

There are other players in the category of special-purpose controllers who compete with the PLCs and controllers that were discussed earlier. The most important players are personal computer-based controllers and programmable automation controllers. These are discussed briefly in the following sections.

13.4.1 PC-Based Controller

The combination of the declining price of commercial personal computers (PCs) and the increasing reliability of operating systems has led to **soft-logic** control applications and the development of PC-based controllers. The benefits of PC-based controller are as follows:

- Cost-effectiveness
- Excellent alternative for many applications
- Networking capabilities

PC-based controllers traditionally link an adapter card on a PC to I/O modules, and they work with customized applications written for control and communication.

Figure 13.15 illustrates the architecture of a PC-based controller. The adapter shown in the figure is the **logical interface** between the PC and the I/O modules over a dual Ethernet interface (see Chapter 14). The communication link can also be a serial interface.

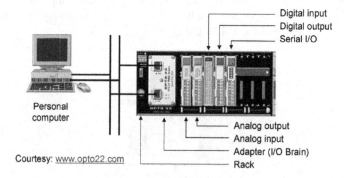

Figure 13.15 PC-based controller.

Figure 13.16 illustrates various functional components associated with the PC-based controller.

13.4.2 Programmable Automation Controller

The programmable automation controller combines the features and capabilities of a PC-based controller and programmable logic controller. Functionally, the PAC and

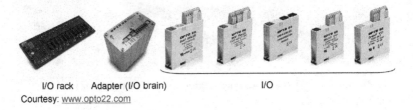

I/O rack   Adapter (I/O brain)   I/O
Courtesy: www.opto22.com

Figure 13.16 Components of PC-based controller.

PLC serve the same purpose, as they are primarily used to perform automation tasks. Compared to PLCs, PACs are open and modular in their architecture. Hence, the hardware selection can be made from the open market without any worry about compatibility. The following are some important features of PACs in comparison with PLCs:

• PACs provide the best of both the PLC and PC worlds.
• PACs are programmed using more generic software tools, while PLCs are programmed in ladder logic diagrams, function block diagrams, etc.
• PACs combine the high performance and I/O capabilities of PLCs with the flexible configuration and integration strengths of PCs.
• PACs have a common hardware platform for wide-ranging purposes.

Figure 13.17 illustrates an industry example of a PAC. Here, the personal computer in the PC-based controller is replaced by a PAC. Also, the adapter (I/O brain) is the logical interface between the processor and the I/O modules.

Figure 13.18 illustrates various functional components associated with a PAC.

In relation to the industry examples previously discussed, the adapter (I/O brain) in a PC-based controller or in a PAC does the following:

• Serves as the interface between a PC and its I/O and PAC
• Manages I/O modules for data acquisition and control
• Can work over an Ethernet I/F or serial I/F (with or without redundancy) for data exchange

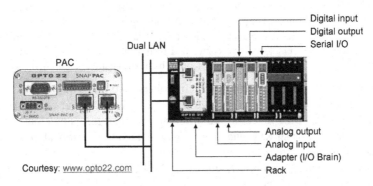

Figure 13.17 Programmable automation controller.

PAC   I/O Rack   Adapter (I/O Brain)   Input/Output
Courtesy: www.opto22.com

Figure 13.18 Components of PAC.

The major difference between a PC-based controller and a PAC is that the PAC employs a **hardened industrial PC** (special PC with high reliability used to work with harsh industrial environments).

13.5 Summary

In this chapter, we discussed the need for special-purpose controllers, namely PLCs, loop controllers, and programmable controllers (controllers). Even though it is technically possible to employ the general-purpose controller for all of the applications, its features are not required in all of the applications. So, the design and development of special-purpose controllers to meet specific application areas led to performance improvement, optimization, and cost-effectiveness.

Today's PLCs, controllers, and RTUs, not differing much in their hardware and software structure, have all the features of general-purpose controllers. They can technically be employed for executing the automation and data communication functions for all types of processes. However, their performance and cost may vary.

Special points to be noted are as follows:

- PLCs are primarily designed and optimized for executing logic functions (sequential control with interlocks) in discrete process automation with speed. They also support continuous process automation in a limited way.
- Controllers are primarily designed and optimized for executing arithmetic/mathematical functions (closed loop control, multi-input and multi-output control) in continuous process automation with accuracy. They also support discrete process automation.
- RTUs are primarily designed and optimized for data acquisition from the process and transmission to the remote control center and receipt of command from remote control center and transfer to process. They also support local automation in a limited way.

With their built-in differences, PLCs, controllers, and RTUs support all the automation functions (data acquisition, data processing, control, and communication). Only the priorities of the functions are different. In practice, it is possible to employ a PLC or controller as an RTU and vice versa, provided the functional, performance, and cost requirements are met.

Finally, this chapter also briefly discussed the PC-based controller and programmable automation controller.

14 System Availability

14.1 Introduction

Any interconnected hardware and software in an automation system, however reliable, are susceptible to failures for a variety of reasons, and this can cause partial or total failure of the automation functions. When a functional unit fails, the corresponding automation functionality becomes unavailable, affecting the overall performance of the industrial process. In some industrial processes, partial or total failure of the automation system may not be a serious issue. In other cases, especially in continuous processes, the automation system needs to function continuously with the utmost reliability and availability. The functioning of the automation system can be a 24×7 job. This chapter addresses the issue of how to design an automation system with 100% availability when required.

We will discuss the provisions for enhancing availability, beginning with some basics on availability enhancement through standby or redundancy.

14.2 Standby Schemes

The following sections discuss various standby schemes, including the no-standby scheme.

14.2.1 No Standby

In the case of failure of a unit (a controller in DCS or FEP/RTU in NCS, its functionality is unavailable for the entire duration of the following sequence:

1. Occurrence of fault
2. Noticing of fault
3. Diagnosing of fault
4. Rectification of fault
5. Reinstallation of rectified unit
6. Recommission of rectified unit
7. Resumption of operation

This sequence is a time-consuming process, and the unit's functionality is unavailable until the completion of the full sequence of actions. Resuming the operation as quickly as possible depends on how well one can reduce the time taken for the sequence.

Overview of Industrial Process Automation. DOI: 10.1016/B978-0-12-415779-8.00014-0

In some cases where the continuous operation of a unit is neither warranted nor demanded, one can wait for the unit to go through the previous sequence of actions. However, this calls for an efficient service setup.

14.2.2   Cold Standby

An extra unit called **cold standby**, which is physically and functionally identical to the working unit, is kept ready for immediate one-to-one replacement if the working unit fails. This way, the failed unit can be repaired without any great urgency. This approach, while providing better availability, leaves a break in the operation of the automation system, and it cannot continue from the state where the working unit had stopped. In other words, the previous data or history stored in the failed unit is not available to the cold-standby unit to continue the function. However, this arrangement is better than no standby, and it is acceptable in many situations, economical, and technically simple.

14.2.3   Hot Standby

The drawback to the cold-standby arrangement (the inability to continue the function from where the operation had stopped) is overcome by the **hot standby**, as illustrated in Figure 14.1. This arrangement provides for a link between the main and the hot-standby unit for the data exchange.

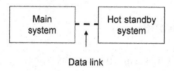

Data link

Figure 14.1 Hot standby with data link.

Normally, the hot-standby unit is kept ready to take over from the failed unit automatically. There are many approaches to the takeover scheme, such as the ones described as follows:

- Both the main and hot-standby units receive the data and update their own databases. Only the designated main unit takes the control actions. As the hot-standby unit is in continuous dialog with the main/working unit, on recognition of any failure, the standby unit automatically takes over the control without any break.
- The hot-standby unit periodically initiates a dialog with the main/working unit, gets the latest data, and updates its own database (data shadowing). In case the main unit does not respond, the standby unit considers the main unit faulty and starts functioning from its latest available data. The maximum loss of data is limited to the time between the two consecutive dialogs (switchover time).

The following sections analyze the availability issues in DCS and NCS and the application of the hot-standby technique to them.

One of the causes of a fatal failure is the failure of the power supply unit in the controller. This can be overcome by having two power supplies in parallel inside the

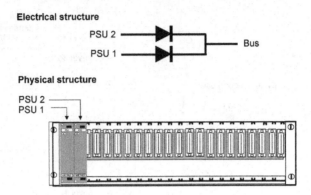

Figure 14.2 Paralleling of power supply units.

rack, as shown in Figure 14.2. Normally, both the power supplies share the load, and upon failure of one of the power supply units, the other healthy one takes the full load. Hence, in the following discussions, power supply units are not considered in the availability analysis and enhancement.

14.3   Distributed Control System

14.3.1   Availability Analysis in DCS

Availability analysis in DCS can be made with the information flow from the process to the instrumentation subsystem (Level 1), to the control subsystem (Level 2), to the communication subsystem (Level 3), and to the human interface subsystem (Level 4). This is illustrated in Figure 14.3. Any failure (full or part of a subsystem) can be fatal or nonfatal. The fatal failure is the one that affects the overall functioning of the DCS. In the following discussions, the failure of a controller or its communication interface, for whatever reasons, leads to the unavailability of automation functions for the subprocess to which the controller is assigned. This is considered to be a major failure.

With the aforementioned information flow, the following sections explain the implications of failure of a particular subsystem or one of its modules at various levels on the availability of automation functions for a subprocess.

14.3.1.1   Level 1: Instrumentation Subsystems

Level 1 has the instrumentation subsystem with the information flow in parallel. It is connected to parallel Input/Output (I/O) channels of the I/O modules in the controller. In other words, one device, for one process parameter, is connected to one I/O channel in one I/O module (see Section 14.5 for I/O redundancy).

14.3.1.1.1   Level 1a: Instrumentation Devices
The failure of any instrumentation device affects only the data acquisition or control function of the particular process parameter connected to the failed device. This

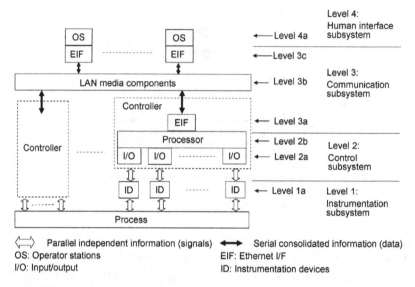

Figure 14.3 Information flow in DCS.

is a nonfatal failure, as it does not totally affect the automation functions of the subprocess.

14.3.1.2   Level 2: Controllers

Level 2 has the control subsystem with two functional modules in series in the information flow, namely, I/O modules and the processor module in the controller.

14.3.1.2.1   Level 2a: I/O Modules

As all I/O modules work in parallel, the failure of any I/O module affects only the data acquisition or control function of the particular process parameters connected to the failed I/O module. This can also be treated as a nonfatal failure, as it does not affect the total automation functions of the subprocess.

14.3.1.2.2   Level 2b: Processor Module

The processor module can fail due to either a hardware failure (power supply, processor, and memory) or a software failure. Processor module failure affects the total controller. This is a fatal failure, as it totally affects the automation functions of the subprocess leading to the unavailability of the following functional subsystems (cascading effect):

- Operator stations dedicated to the failed controller, even though they may be healthy
- Other operator stations in DCS that need interaction with the failed controller
- Other controllers in DCS that need interaction with the failed controller

14.3.1.3   Level 3: Local Communication Subsystem

Level 3 has the communication subsystem with three functional modules in series in the information flow, namely the Ethernet I/F in the controller, local communication network, and Ethernet I/F in the operator station.

14.3.1.3.1   Level 3a: Ethernet I/F in Controller

With the failure of its Ethernet I/F, the controller cannot communicate with the subsystems at the same level (other controllers) and at the higher level (operator stations). This makes the operator interaction totally unavailable (similar to the failure of the processor module).

However, with a healthy processor module and I/O modules in the controller as well as healthy instrumentation devices, the execution of the automation functions of the assigned subprocess is not affected—except where the controller is dependent on the operator stations and other controllers. Hence, this failure can be treated as nonfatal, as it does not affect the automation functions of the subprocess.

14.3.1.3.2   Level 3b: LAN Media Components

The LAN media components (cables, connectors, and hub), being passive, generally do not contribute to the failures of the network. Therefore, this is not considered in the availability analysis.

14.3.1.3.3   Level 3c: Ethernet I/F Module in Operator Station

This failure makes the concerned operator station totally unavailable (cannot interact with the controllers); even though the other functional parts of the operator station may be healthy. This amounts to the failure of the operator station discussed in the next section. Here also, the failure is nonfatal, as we miss only the human interface functions assigned to the failed operator station, which does not affect the functioning of the automation of the subprocess.

14.3.1.4   Level 4: Operator Station

Level 4 at the apex has a computer system functioning as the operator station and is built of hardware (power supply, processor, memory) and software (human interface functions).

14.3.1.4.1   Level 4a: Operator Station

The failure of an operator station can be due to the failure of its computer system (hardware and/or software) or of its Ethernet I/F module (operator station becoming totally ineffective). However, whether the failure is due to its Ethernet I/F or the computer system, the human interface functions of the failed operator station can be moved to one of the healthy operator stations. Hence, this is also a nonfatal failure, as it does not affect the automation of the concerned subprocess. However, the only minor drawback of this approach is the extra loading of one of the operators of the healthy operator stations.

14.3.2   Availability Enhancement in DCS

Next, we will discuss availability enhancement in DCS.

From the preceding sections, we can conclude that the most critical modules in the information flow whose individual failure contributes to the fatal failure in DCS are the following:

- Processor in controller
- Ethernet I/F in controller
- LAN

With the assumption that only one critical module in the information flow can fail at any given time, a way to increase the availability of the overall DCS functioning is to provide for a redundancy, or hot standby, to these critical modules.

14.3.2.1   Processor in Controller

Figure 14.4 illustrates the standby arrangement for the processor module.

The hot-standby processor automatically takes over the function of the main processor whenever it fails.

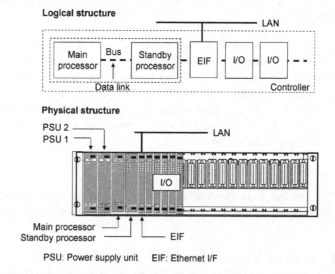

Figure 14.4 Standby for processor in controller.

14.3.2.2   Ethernet I/F in Controller

Failure of the controller's Ethernet I/F can lead to the isolation of the controller with other subsystems (other controllers, operator stations, etc.) on the LAN. This fatal failure can be overcome by providing standby/redundancy to the Ethernet I/F, as shown in Figure 14.5, with the Ethernet I/F integrated with the processor module.

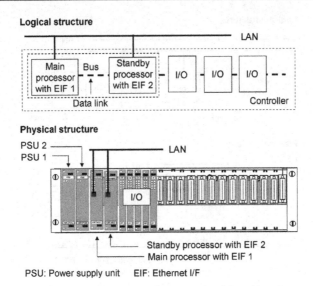

Figure 14.5 Standby for processor and Ethernet I/F in controller.

On recognition of the failure of the Ethernet I/F in a processor module, the other processor module with a healthy Ethernet I/F takes over and continues the communication over LAN.

14.3.2.3 LAN

In order to further increase the availability of the local communication systems, one can consider the physical redundancy for LAN components, as shown in Figure 14.6. This arrangement also addresses the possible failure of any LAN component.

In this arrangement, one Ethernet I/F is connected to LAN 1, while the other is connected to LAN 2. Simultaneous failure of Ethernet I/F 1 and LAN 2 or Ethernet I/F 2 and LAN 1 does lead to total failure irrespective of duplication of the processor and Ethernet I/F in the controller and the LAN.

To overcome this drawback, the arrangement employs main and hot-standby processors with two Ethernet interfaces each. These are connected to both the LANs. Any simultaneous failure of Ethernet I/F 1 and LAN 2 or Ethernet I/F 2 and LAN 1 does not lead to the total failure of the local communication. This arrangement is illustrated in Figure 14.7.

14.4 Network Control System

Most of the discussion about DCS holds true for NCS as well, especially at the top three levels (OS, LAN, and FEP). The following sections discuss the availability analysis and enhancement at other levels.

Logical structure

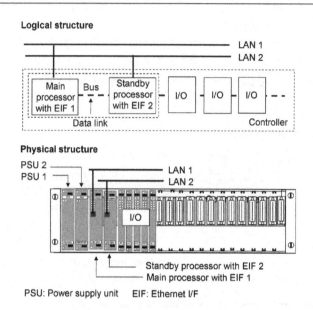

Figure 14.6 Standby for LAN media components.

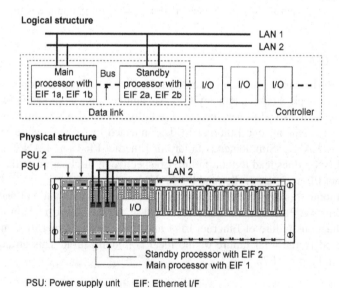

Figure 14.7 Dual LAN with dual Ethernet I/F in controller.

14.4.1 Availability Analysis in NCS

Availability analysis in NCS can be made with the information flow in a distributed process from process to instrumentation subsystem (Level 1), to control subsystem

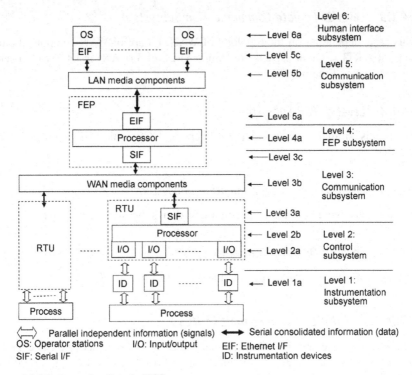

Figure 14.8 Information flow in NCS.

(Level 2), to remote communication subsystem (Level 3), to FEP (Level 4), to local communication subsystem (Level 5), and to human interface subsystem (Level 6). This is shown in Figure 14.8. As in the case of DCS, any failure (full or part of a subsystem) in the information flow can be fatal or nonfatal. The fatal failure is the one that affects the overall functioning of the NCS. In the following discussions, the failure of an RTU or its communication interface, for whatever reasons, leads to the unavailability of automation functions for the subprocess to which the RTU is assigned. This can be considered as acceptable.

Using this information flow, the following sections illustrate the implications on automation functions of the failure of a particular subsystem or one of its modules at various levels.

14.4.1.1 Level 1: Instrumentation Subsystem

This is identical to Level 1 in DCS with RTU in place of the controller.

14.4.1.2 Level 2: RTUs

This is also identical to Level 2 (controller) in DCS with RTU in place of the controller.

14.4.1.3   Level 3: Remote Communication Subsystem

In this case, depending on the topology of WAN, failure of WAN components may affect a few RTUs. Other aspects are similar to Level 3 (LAN) in DCS with a serial I/F in place of an Ethernet I/F.

14.4.1.4   Level 4: FEP Subsystem

This case is similar to Level 2 (controller) in DCS with FEP in place of the controller.

14.4.1.5   Level 5: Local Communication Subsystem

This case is also similar to the local communication subsystem (LAN) in DCS.

14.4.1.6   Level 6: Operator Station

This case is similar to Level 4 (human interface subsystem) in DCS.

14.4.2   Availability Enhancement in NCS

Availability enhancement in NCS is similar to that of DCS. Hence, we can conclude that the most critical modules in the information flow whose individual failure contributes to the fatal failure in NCS are the following:

- Serial I/F and processor in RTU
- WAN
- Serial I/F, processor, and Ethernet I/F in FEP
- LAN

Once again, the assumption here is that only one critical module in the information flow can fail at any given time. The approach is to increase the availability of the overall functioning of the NCS to provide for a redundancy or standby to the critical module.

14.4.2.1   Serial I/F and Processor in RTU

Figure 14.9 illustrates the redundancy arrangement in RTU for both serial I/F and processor.

Here, the hot-standby processor takes over the functions of the failed main processor with data updating over the RTU bus.

14.4.2.2   WAN

Figure 14.10 illustrates the redundancy for WAN.

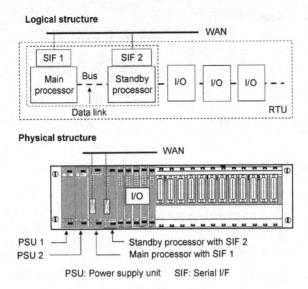

Figure 14.9 Standby for processor and serial I/F in RTU.

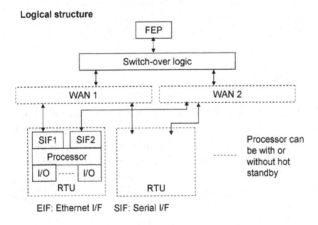

Figure 14.10 Redundancy for WAN.

14.4.2.3 Serial I/F, Processor, and Ethernet I/F in FEP

Figure 14.11 illustrates the provision of standby at Ethernet I/F, processor, and serial I/F in FEP. LAN itself is used as the data link between the main processor and the hot-standby processor.

14.4.2.4 LAN

LAN redundancy is identical to LAN redundancy in DCS.

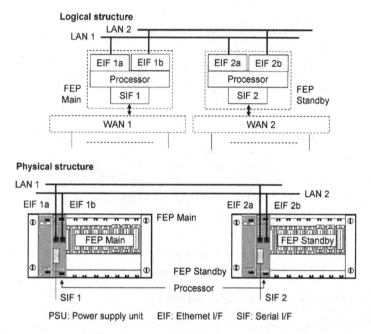

Figure 14.11 Standby for FEP processor and communication interfaces.

14.5 I/O Redundancy

So far, the application of standby/redundancy concepts for vital components in the automation systems has been discussed. Improving the availability in the automation system is meaningful provided there is adequate redundancy in the process plant for its vital equipment. This is not discussed here as it is out of scope. However, to support redundant equipment in the process plant, an automation system can support I/O redundancy for the redundant plant equipment with its associated instrumentation.

14.6 Summary

In this chapter, we discussed the issues related to the availability of DCS and NCS when a specific critical module in the information flow fails. On recognition of the failure of the main unit, several types of standby schemes may be employed. Hot standby is utilized to continue the operation from the point where the main unit failed. However, as this approach is complex and expensive, it is implemented only where absolutely required.

15 Common Configurations

15.1 Introduction

In Chapter 14, we discussed how to provide hot standby for various critical components in the information flow link so that the total operation of the automation system, be it distributed control system (DCS) or network control system (NCS), is always available. In this chapter, using the approaches available for enhancing system availability, some common configurations are discussed for increase in functionality of both DCS and NCS.

15.2 Distributed Control System

In the following sections, some typical configurations of the DCS above the controller level are discussed, assuming that the controllers are already provided with hot standby.

15.2.1 Operator Stations

Figure 15.1 illustrates the configuration of the controllers with hot standby. This is the basic configuration wherein one or more operator stations with a specific area or function assignment carry out the operator functions. Normally, any operator station can be used for monitoring the process variables assigned to the other operator stations. However, the issue of commands to the process is restricted only to the authorized operator station. If one of the operator stations fails, another healthy operator station, under authorization, can perform its full functions. The drawback of this approach is that the operator station to which the functions of the failed operator station are transferred gets overloaded.

15.2.2 Supervisory Stations

In many applications, there are some common or supervisory functions for coordinating the functions of the controllers. It is generally not preferable to share one of the controllers for this function, as the performance of the controllers gets reduced. If we incorporate supervisory functions in any operator station, it becomes unavailable if that particular operator station fails. To overcome these problems, the supervisory functions are provided on an independent supervisory station, as illustrated in Figure 15.2. The

Overview of Industrial Process Automation. DOI: 10.1016/B978-0-12-415779-8.00015-2

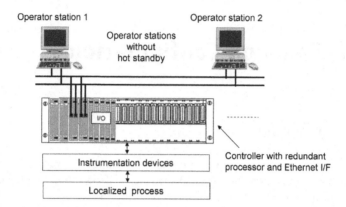

Figure 15.1 Operator stations without standby.

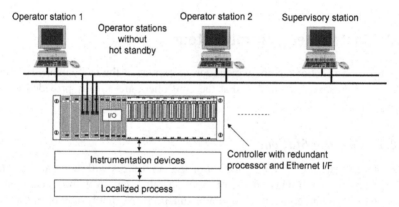

Figure 15.2 Supervisory station without standby.

independent station for common supervisory functions is called a server, and the present day servers hold the complete process data unlike the controllers of the past.

If the supervisory functions are critical, the obvious next step is to provide a hot standby for the supervisory station, as illustrated in Figure 15.3.

15.2.3   Application Stations

In some special cases, there is a need to execute complex real-time application functions with heavy computations, and these must be performed within a fixed time. The combining of complex application functions with supervisory functions may lead to the performance degradation of both functions. Once again, to overcome this, an independent station (with or without hot standby) can be employed exclusively

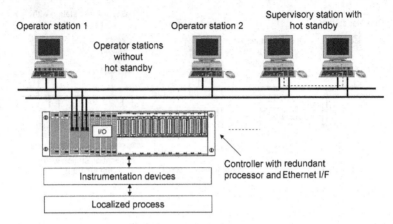

Figure 15.3 Supervisory station with standby.

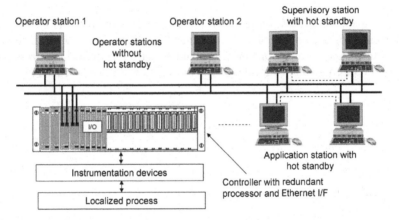

Figure 15.4 Application station with standby.

for the execution of the complex application programs. Figure 15.4 illustrates the configuration of an independent application station with hot standby. These independent stations are called application servers.

15.3   Network Control System

All the previous discussions pertaining to DCS are equally applicable to NCS. As shown in Figure 15.5, the only difference is that front-end processors (FEPs) replace the controller level in NCS.

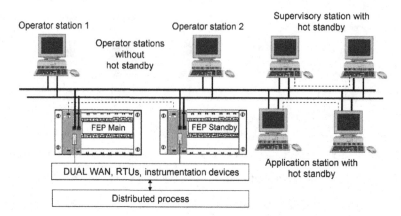

Figure 15.5 Supervisory and application stations with standby.

15.4 Summary

In this chapter, we discussed a few typical configurations of DCS and NCS, such as basic monitoring and control of the localized and distributed processes and the execution of complex application functions. It should be noted that the technical complexity and cost of the configuration increases with the addition of more functionalities.

16 Advanced Input/Output System

16.1 Introduction

In Chapter 8, we discussed the construction of general-purpose controllers with multiple communication modules, as illustrated in Figure 16.1. In all the discussions, we selected a configuration of the controller where the input/output (I/O) modules are placed in the same rack as the processor module. Here, the I/O modules are close to the processor physically and logically, and they adopt parallel data transfer with the processor over the bus. In case of more I/O modules (which are application-dependent), additional or supplementary racks are employed with the bus physically extended. This I/O arrangement is called centralized I/O (CIO).

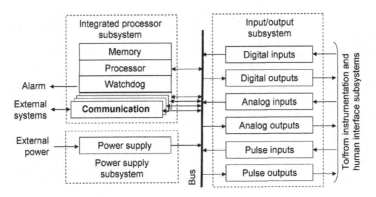

Figure 16.1 Controller with centralized I/O.

16.2 Centralized I/O

In CIO, the I/O is physically and logically an integral part of the controller, and the instrumentation devices are interfaced to the I/O channels on a one-to-one or hardware basis. Each process signal requires one exclusive I/O channel for information transfer in one direction, depending on whether the signal is an input or an output. Further, the number of instrumentation devices required equals the number of process inputs and outputs. The information transferred between the controller and the instrumentation devices is in the form of an **electronic signal** representing the physical signal.

The physical interconnections between the instrumentation devices and the CIO are illustrated in Figure 16.2.

Overview of Industrial Process Automation. DOI: 10.1016/B978-0-12-415779-8.00016-4

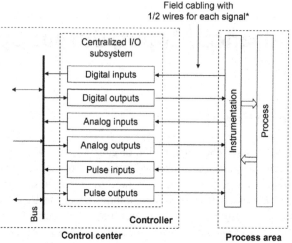

Figure 16.2 Field cabling in CIO.

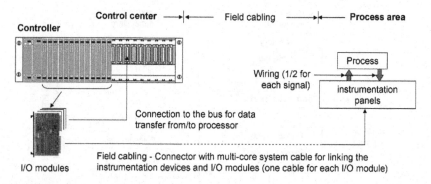

Figure 16.3 CIO engineering.

As a part of the controller, the CIO is installed within the control room and is generally away from the instrumentation devices and the process. Figure 16.3 illustrates the engineering and physical interconnections between the instrumentation devices and the CIO. The number of wires in the system cable depends on whether the signals are isolated or not.

16.2.1 Intelligent CIO

Traditionally, I/O modules were not intelligent and were instructed to perform their functions by the processor. The processor had to scan the channels in all input modules for new data, processing, diagnostics, etc. This resulted in too much computational burden on the processor, leading to a reduction in the controller's overall performance. Making the I/O modules intelligent (microprocessor-based) was not feasible earlier in

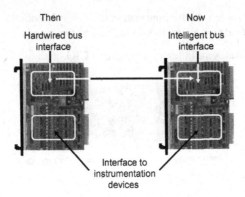

Figure 16.4 Transition to intelligent I/O.

view of the cost considerations. Cost-effective microprocessors and microcontrollers have now made intelligent I/O a reality.

Physically, there is no difference between non-intelligent and intelligent I/O modules, and they are one-to-one replaceable, as illustrated in Figure 16.4. Changes are required only in the software. However, the intelligent I/O contributes to the overall improvement in the performance of the controller by taking away the routine I/O data processing load from the processor. The performance improvement generally comes from the following:

- Processing of the acquired data from the process in the I/O module itself before sending it to the processor (*analog inputs*: filtering, linearization, offset correction, limit checking, etc.; *digital inputs*: anti-bouncing, time tagging, sequencing, etc.)
- Processing of commands received from the processor in the I/O module itself before sending it to the process (*analog outputs*: set-point manipulation, etc.; *digital outputs*: command security checking, command execution monitoring, etc.)
- Execution of diagnostic program of the I/O module in the I/O module itself and reporting any faults to the processor

16.2.2   Advantages and Disadvantages

The advantages of CIO are the following:

- Ideal for small and less spread out processes.
- Controller in control center is less expensive, as I/O modules are placed close to the processor.
- Fast data transfer between the processor and I/O modules due to parallel communication over the bus.

The disadvantages of CIO are the following:

- Controller is bulky, occupies more space, and consumes more power because the I/O is concentrated.
- High cost of control cabling between the instrumentation devices and the controller (including the installation and maintenance).

- **Signal transfer** between the instrumentation devices and controller (possibility of information quality degradation and/or loss during information transfer)

Figure 16.5 illustrates an industry example of CIO.

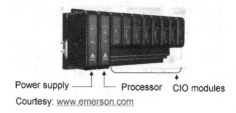

Power supply ⎯⎐   ⎐⎯ Processor   CIO modules
Courtesy: www.emerson.com

Figure 16.5 Centralized I/O.

16.3   Remote I/O

The biggest disadvantages of CIO, cables and cabling cost, led to the evolution of remote I/O (RIO), a special-purpose controller, which exploited the communicability feature of the controller. It moved I/O closer to the process, reducing the cabling cost to a minimum, as illustrated in Figure 16.6.

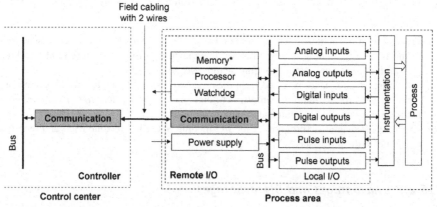

*Program for data acquisition from the process/transfer to controller and receipt of data from the controller/transfer to process.

Figure 16.6 RIO structure.

As seen here, the RIO is not physically an integral part of the controller, and the interfacing between the I/O and the controller is over a pair of serial communication modules. The features of RIO are the following:

- Single pair of signal cables for bidirectional information transfer of many process parameters over a pair of communication interface modules
- **Data transfer** (not signal) between the controller and RIO (no possibility of information degradation and/or loss during information transfer)

Figure 16.7 illustrates the physical structure of an industry example of RIO.

Figure 16.8 illustrates multiple RIOs installed in the field (closer to the process) on a single serial communication cable in multidrop mode, which is several units sharing the same serial link.

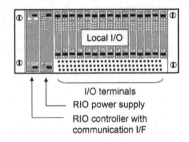

Figure 16.7 Remote I/O.

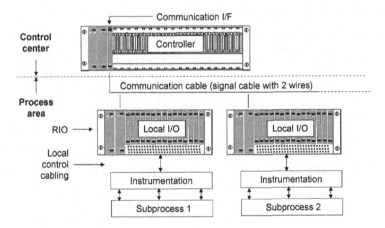

Figure 16.8 Distributed RIO.

The I/O that existed earlier in the control center as an integral part of the controller is relocated to the process area to become an integral part of RIO, reducing the cabling cost drastically. Physically, RIO resembles the controller, but it has a specific function. RIO just acquires the data from the process through its local input modules and dispatches it to the controller after some local processing. Similarly, in the other direction, RIO receives the data from the controller and dispatches it through its local output modules to the process after some local processing.

Figure 16.9 illustrates the engineering and physical interconnections between the instrumentation devices and the RIO.

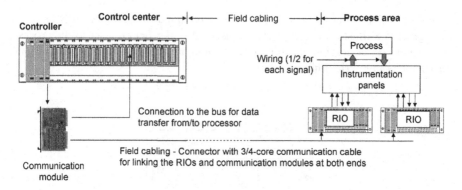

Figure 16.9 RIO engineering.

16.3.1 Advantages and Disadvantages

The advantages of RIO are as follows:

- Suitable for large and physically spread out processes
- Controller in control center is slim and power-efficient because the I/O is distributed
- Much less cabling between process instruments and controller (two wires linking all the RIOs) result in less installation and maintenance cost
- Data transfer (not signal) between the RIO and controller (no signal quality degradation)

The disadvantages of RIO are as follows:

- Expensive, as they are intelligent and communicable
- Slow data transfer between the controller and RIO modules due to serial communication with the controller—not as much of an issue now considering the faster response in present day communication

Figure 16.10 illustrates an industry example of RIO.

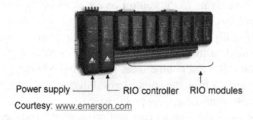

Power supply ┘ └ RIO controller   RIO modules
Courtesy: www.emerson.com

Figure 16.10 Remote I/O—Industry example.

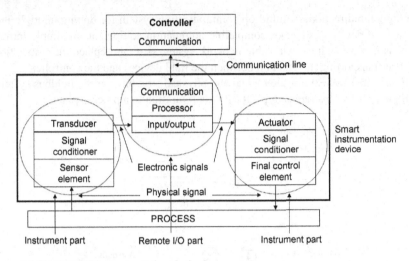

Figure 16.11 SMART instrumentation device.

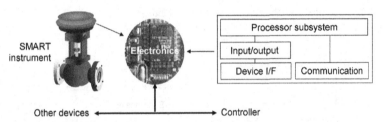

Communicable data acquisition and control unit with interface to the instrument (sensor, signal conditioning, and transducer) and to communication line - All integrated and embedded

Figure 16.12 SMART device—construction.

16.4   Fieldbus I/O

The obvious next step is to merge the RIO and the traditional instrumentation devices to form intelligent and communicable instrumentation devices or **self-monitoring analysis reporting technology (SMART)**[1] devices, as illustrated in Figure 16.11.

Figure 16.12 illustrates the construction of the SMART instrumentation device.

The fieldbus is a serial bus that runs throughout the process plant, linking all the smart I/O devices with the controller. The communication module is called the fieldbus I/F, the I/O system is called the fieldbus I/O (FIO), and the SMART device is called the fieldbus device (FD).

The conventional way has been hardwiring (cabling) to carry a single process variable (input or output) either in discrete or in analog form. On the contrary, the fieldbus

[1] InTech, ISA, March 2006.

is a digital, bidirectional, multidrop communication system for linking multiple intelligent field devices and other compatible automation equipment. In simple terms, a fieldbus is a two-wire single cable network system that can replace the conventional ways of transporting information between the field devices and the controller.

Figure 16.13 illustrates the logical and physical structures of the fieldbus system. Figure 16.14 illustrates a typical FIO configuration in a process plant.

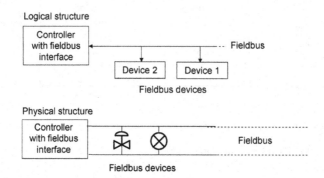

Figure 16.13 Logical and physical structure of fieldbus.

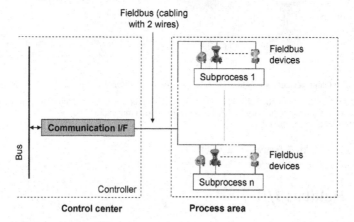

Figure 16.14 FIO configuration.

16.4.1   Advantages and Disadvantages

The advantages of FIO are the following:

- It has all the same advantages as RIO.
- Fieldbus devices, being intelligent, can diagnose their own health and send the diagnostic information to the controller for proactive maintenance.
- Fieldbus devices can perform local processing of data themselves, reducing the burden on the controller.
- Fieldbus devices can implement specific automation functions locally (for example single loop control), hitherto centrally executed in the controller. This not only improves the

performance of the controller but also eliminates the timing constraints/conflicts in the execution of multiloop functions with different priorities.
- Unlike traditional instrumentation devices, which provide one type of data per device, the fieldbus devices provide almost unlimited information (diagnostics, status, etc.) about themselves to the controller and to all the other members on the network.
- Facilitates bidirectionality, remote calibration, field-based control, plug and play, and high-speed connection.
- It has data integrity to send/receive only good data, so no data is lost. The controller or a fieldbus device can request retransmission in case bad data is received.
- Enables seamless integration of information right from the field level to the management level.

The disadvantages of FIO are the following:

- It has all the disadvantages of RIO.
- Fieldbus devices are relatively expensive and complex.
- Presence of mixed signals on the same bus.
- The architecture, configuration, monitoring, diagnostics, and maintenance tools are highly complicated.

Figure 16.15 illustrates som industry examples of fieldbus devices.

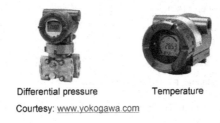

Differential pressure   Temperature
Courtesy: www.yokogawa.com

Figure 16.15 FIO Fieldbus devices.

Figure 16.16 summarizes the advantages of FIO over CIO.

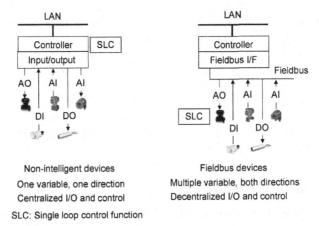

Figure 16.16 Comparison of CIO and FIO.

Figure 16.17 illustrates an example of an intelligent and communicable energy meter employed in the electrical industry for the measurement of all electrical parameters of an outgoing feeder. In this case, the traditional approach is to have independent transducers for each power system parameter, which requires either feeder voltage input or feeder current input or both. The intelligent meter/transducer takes feeder voltage and feeder current inputs and computes all the parameters, then sends them to the controller.

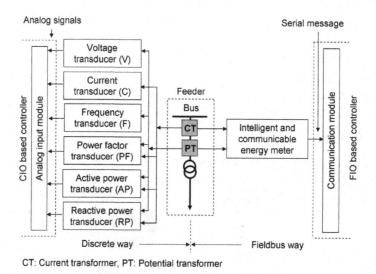

Figure 16.17 Discrete and intelligent transducers.

Figure 16.18 illustrates the construction and functioning of an intelligent and communicable energy meter, or transducer. Here, the waveforms of voltage (PT) and current (CT) inputs are continuously captured, and the instantaneous values of all the electrical parameters are computed as functions of the input voltage and current and sent to the controller on a serial link as digital data. The energy meter, being intelligent and communicable, can compute lots of data, such as energy, maximum demand, and send this to the controller. In this way, a single intelligent and communicable device can replaced many non-intelligent devices.

The fieldbus devices in process industry (temperature, level, pressure, etc.) physically look like their non-intelligent counterparts even though they are totally different functionally.

16.4.2   Fieldbus I/F Module

The serial I/F employed in FIO is called **fieldbus I/F**. So far, we have been talking about communication programs or software that facilitates the information transfer

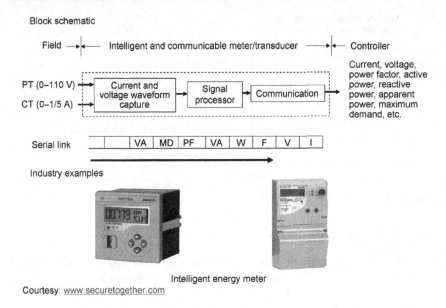

Figure 16.18 Intelligent energy meter—industry examples.

between any two intelligent devices with the help of interfaces present at both ends and supported by an appropriate communication program. The participating devices are controller and RIO/fieldbus devices. The communication between any two intelligent devices can take place only if they are compatible in their connectivity, in terms of both hardware and software. This compatibility in connectivity is called **communication protocol**. This is a formal set of rules and conventions governing the hardware and software connections for information exchange among two or more intelligent systems.

Figure 16.19 illustrates the traditional way of communication using a hardware or non-intelligent communication interface (communication module with serial interfacing).

The communication protocol, which uses special software, resides in the controller memory, and it is executed by the controller (in addition to the execution of automation functions). The conventional non-intelligent serial interface module facilitates the following:

- Establishing of the **hardware connectivity** between the partners
- Assembly of serial information received from the sender into parallel information (serial to parallel conversion) and transfer of this to the processor over the bus
- De-assembly of parallel information transferred from the processor over the bus into serial information (parallel to serial conversion) and sending this to the receiver over the medium

Because of its high overhead, the execution of the communication protocol software demands considerable processor time from the controller. This communication is beyond just the normal data transfer between the two partners and is extensive, continuous, and repetitive. This can drastically reduce the performance of the controller.

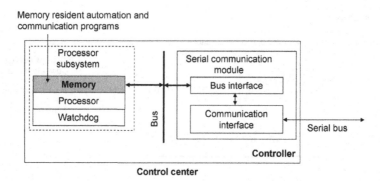

Figure 16.19 Non-intelligent serial I/F.

16.4.3   Intelligent Serial I/F

The shortcomings in the traditional non-intelligent serial I/F led to the development of the intelligent serial I/F, which decentralized the execution of automation and communication functions, as illustrated in Figure 16.20.

Here, the local processor in the intelligent serial I/F module takes away the communication load from the controller.

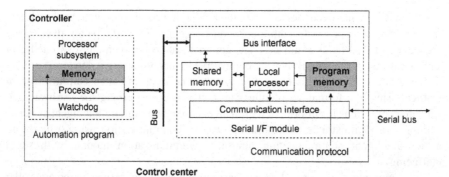

Figure 16.20 Intelligent serial interface.

16.4.4   Protocol Standards

The most commonly used hardware connectivity standards are **RS-232c** and **RS-485** for serial interfaces and **RJ-45** for Ethernet interfaces, established by standards organizations. Similarly, there are many industry standards for fieldbus **software connectivity**. Some of the important examples are as follows:

- Actuator–sensor interface,[2] Interbus,[3] Ethercat,[4] etc., for discrete process automation

[2] http://as-interface.net.
[3] http://www.interbusclub.com.
[4] http://www.ethercat.org.

- Profibus,[5] Foundation,[6] HART,[7] Modbus,[8] etc., for continuous process automation
- LON[9] for building automation and industrial data networking

Differential pressure   Temperature

Courtesy: www.yokogawa.com

Figure 16.21 Fieldbus devices with wireless communication.

This communication protocol software resides in all the partners for information transfer. More standards are specified by IEC 61158. This standard included several established protocols. Today, the communication protocol generally refers to the software connectivity, as the hardware connectivity standards are frozen. The intelligent serial I/F module can be personalized to support different protocols by just changing the protocol software in the module.

Figure 16.21 illustrates an industry example of a fieldbus device with wireless communication.

16.5 Summary

In practice, the green field projects can go directly to FIO. However, in case of expansion or retrofitting of a plant, the existing, or legacy instrumentation devices can be made part of FIO using RIO to support them. This approach protects the investment already made on legacy devices. Both RIO and FIO are required to support the common communication protocol. Figure 16.22 illustrates the transition of the I/O system from CIO to RIO to FIO.

The fieldbus technology facilitates decentralization of intelligence in the automation system through a network of controller and fieldbus devices.

---

[5] http://www.profibus.com.
[6] http://www.fieldbus.org.
[7] http://www.hartcomm.org.
[8] http://www.modbus.org
[9] http://www.echelon.com.

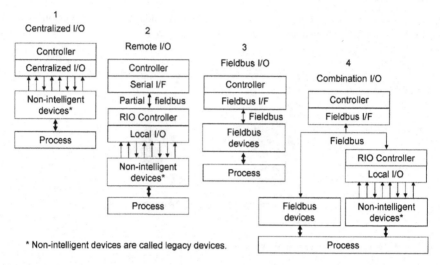

Figure 16.22 Transition in I/O subsystem.

17 Concluding Remarks

17.1 Introduction

We have now completed a tour of the basics of industrial process automation. In addition to summarizing the major functions of automation, this concluding chapter will briefly discuss the trends in automation.

17.2 Major Functionalities

The following sections summarize the major functions of automation systems discussed in the book.

17.2.1 Data Acquisition

The primary function of the automation system is to acquire the process data in the control center without which no further actions are possible. This data may come from a local source or remote source. The controllers in centralized control system (CCS) or distributed control system (DCS) acquire the local data, while remote terminal units (RTUs) in network control system (NCS) acquire the remote data. Typical examples of process data are the values of continuous/analog parameters and the states of discrete/digital parameters.

17.2.2 Data Supervision or Monitoring

Once the process data is available in the control center, the next job of the automation system is to supervise or monitor the data for any deviation from its normal value or state and to generate an audio–visual alarm to alert the operator to take appropriate actions immediately. The system handles process alarms/events and system alarms/events.

17.2.3 Process Survey

All of the process information, available in the control center, can be called up on the screens of operator stations for an entire process survey by the operator (with no need to visit the process). The operator can see the current happenings in the process whether the process is local or distributed (even geographically). Typical examples

Overview of Industrial Process Automation. DOI: 10.1016/B978-0-12-415779-8.00017-6

are the chemical reactions in a distillation column or the line loading conditions in an electrical transmission grid.

17.2.4 Process Control

Process control, or the operation in the plant to change the value or state of process parameters, is managed by the control center through operator stations. The devices can be in a localized process or in a distributed process. Typical examples are the opening or closing of a valve in a demineralization plant or the changing of the set-point value of a speed control system in a remote power plant.

17.2.5 Process Studies

The basic online functions of automation systems are process data acquisition, process data analysis/decisionmaking, and process control. These are not enough to produce the optimum results always and calls for running special application programs. These programs optimize the control inputs to the process based on the information available in the control center. Typical examples of process studies are optimization of coal firing for maximum heat generation in a boiler or management of electrical transmission and distribution for maintaining grid frequency and voltage.

17.2.6 Human Interaction

Operator stations provide for human interaction with the process not only for observing the values and states of process parameters but also for effecting manual controls and actions. In fact, the operator station is the tool for total management of the process. Today's operator stations can be in remote places or even Web-based.

17.2.7 Data Logging and History Generation

Data logging stations produce on-demand/periodic logs and reports on hard copies. The automation system also generates process history for future use. Data logging and history generation need not always make hard copies. The data can be logged on different media or systems for off-line analysis at a later date.

17.2.8 Data Exchange

Automation systems are communicable, meaning they can exchange data with other compatible systems either locally or remotely.

17.3 Data Availability

Automation systems are computer based, so they can receive and store a large amount of data in the control center. All of the data received and stored is not required for normal

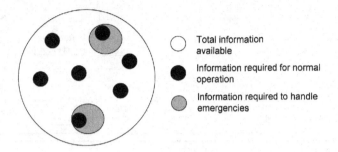

Figure 17.1 Data availability and usage in automation system.

operations. Only very limited data on the object of interest is required to handle the object. However, in case of any abnormality in the functioning of the object, a little more data related to the object is required to handle the situation. The power of the processor/computer-based system is that all of the data is available at the operator's fingertips so that any contingency can be handled.

Figure 17.1 illustrates the amount of data available in the automation system control center and the data in pockets required to handle normal and abnormal situations in a specific situation.

17.4   Today's Automation Systems

Figure 17.2 illustrates the extension of the basic automation system (as shown in the shaded area in the figure) to cover more applications.

There are several additional features in today's automation system, and they include the following:

- *Servers*: These are the heart of automation system, and they not only store the real-time data but also coordinate the activities of all other subsystems. Servers normally have redundancy.
- *Remote clients*: These are operator stations that are located outside the plant and connected to the automation system in the plant.
- *History server*: It is a place where the history (alarm data, event data, process data, etc.) is stored for off-line analysis.
- *Enterprise resource planning (ERP) system*: An ERP is an arrangement for integration and management of information across the entire organization. It covers all business activities, such as finance/accounting, sales, manufacturing, supply, and service.

The major advantage of the previous arrangements is that they present a network for **seamless information flow** from the field level to the management level. This helps to implement the management decisions at the plant/field level automatically. Earlier, automation systems were isolated (islands) from management information systems and the link was manual.

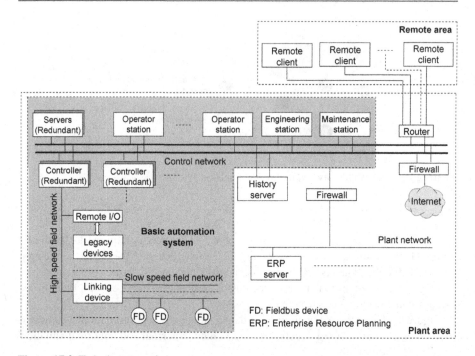

Figure 17.2 Today's automation system.

17.5 Modern Control Center

In the modern control center, whenever there is disturbance in the process (alarms, failures, etc.), the automation system generates an audio alarm prompting the operator to examine the details on display screens and take appropriate and immediate actions. Moreover, modern control centers are designed **ergonomically** to make the control center environment fit for operator comfort, as they have to work for long periods with repetitive activities. Modern control centers are generally, but not solely, equipped with the following:

• Operator stations for manual interaction with the process
• Telephone facilities for interacting with the plant and outside personnel
• Control desks to house the operator stations and telephone equipment
• Large screen displays for real-time display of vital process information
• Printers, plotters, etc., for taking hard copy of the information

Figure 17.3 illustrates an industry examples of a modern control center.

17.6 Application Areas of Automation Systems

In this book, we discussed three kinds of control—discrete, continuous, and hybrid—and their application areas. These are revisited here.

Courtesy: www.emerson.com

Figure 17.3 Modern control center.

17.6.1 Discrete Process Automation

Discrete process automation is further classified as the following:

- General purpose
- Manufacturing

    · Machine tools
    · Motion control
    · Robotics

17.6.2 Continuous Process Automation

Continuous process automation is further classified as the following:

- General purpose
- Process industries

    · Chemical
    · Petrochemical
    · Metal
    · Power generation

17.6.3 Batch Process Automation

So far, we discussed the automation systems designed for particular processes. However, there are automation systems that can be used for several similar processes in batches by just changing the automation program as you would change a recipe. These processes are called batch processes.

Batch process automation is normally employed in life sciences or biotechnology industries, such as pharma, food, and beverage. The batch process can be discrete, continuous, or a combination.

17.7 Summary

After your guided tour on industrial process automation, this concluding chapter summarizes the overall functions of the automation system, with specific reference to data handling and classification based on the applications. It concludes with a discussion of the extension of traditional automation systems to cover additional functions for seamless integration of data from field level to management level and modern control centers.

Appendix A
Hardwired Control Subsystem

A.1 Introduction

In the early days, automation systems were purely based on pneumatic, mechanical, and hydraulic components for both measurement and control. All of these were slow, bulky, sluggish, and less reliable, and they required more maintenance, space, and power. Over time, automation technology moved to electrical, electronic, and finally processor-based/information technology. Advances in electronics, information, communication, and networking technologies played a vital role in making the entire control subsystem more compact, power-efficient, reliable, flexible, communicable, and self-supervisable.

Control subsystems started out with hardwired systems in line with the technology available at the time. This appendix details the beginning of modern control subsystem implementation, which was mainly hardware-based or hardwired for both discrete and continuous processes.

Throughout our discussion of the implementation of various technologies of control subsystems, we will employ the example of a water heating process with its instrumentation and human interface subsystems.

A.2 Discrete Control

As discussed in Chapter 6, discrete process automation systems (open and sequential control with interlocks) handle only the discrete inputs and outputs from the instrumentation and human interface subsystems.

A.2.1 Relay Technology

Modern programmable control systems started with relay technology for automating the assembly lines in automobile industries. In this technology, the simple electromechanical relay, as illustrated in Figure A.1, is the central control element for implementation of the automation strategy.

A simple electromechanical relay consists of a coil of wire surrounding a soft iron core (electromagnet), an iron yoke to provide a low reluctance path for magnetic

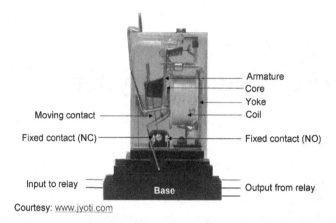

Courtesy: www.jyoti.com

Figure A.1 Electromechanical relay.

flux, a movable iron armature, and a pair of contacts. The armature hinged to the yoke is mechanically linked to a moving contact, and it is held in place by a spring action. In the de-energized state, there is an air gap in the relay magnetic circuit leaving the armature in its normal state. On the contrary, in the energized state, the air gap in the relay magnetic circuit gets closed, moving the armature to its forced state. The armature stays in the forced state as long as the relay remains energized.

Depending on the mechanical arrangements of the contacts, the de-energized state of the relay is called the normal state, which can either have the contacts open (normally open—NO) or the contacts closed (normally closed—NC).

Figure A.2 illustrates the mechanical arrangement of two contacts in NO and NC relays. Normally, under de-energized conditions, the contacts remain open in NO relays and remain closed in NC relays. When the relay is energized, the mechanical action closes the contacts in NO relays allowing the electrical signal to pass through,

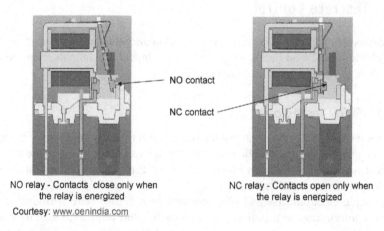

NO relay - Contacts close only when   NC relay - Contacts open only when
the relay is energized   the relay is energized

Courtesy: www.oenindia.com

Figure A.2 Construction of NO and NC relays.

and it opens the contacts in NC relays not allowing the electrical signal to pass through. *NO relays function as discrete signal buffers or signal repeaters while NC relays function as signal inverters.* These relays basically **repeat** the discrete/digital input signal with or without inversion, depending on whether the relay is NO or NC.

A general-purpose relay with three contacts, called a changeover (CO) relay is illustrated in Figure A.3. CO relays are flexible, as they can be wired either as NO or NC relays.

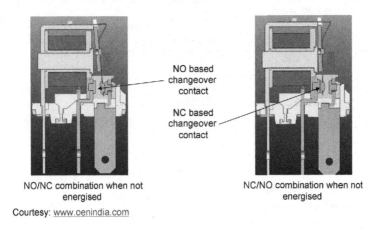

NO based changeover contact

NC based changeover contact

NO/NC combination when not energised

NC/NO combination when not energised

Courtesy: www.oenindia.com

Figure A.3 Construction of CO relay.

There are also relays with additional/parallel NO, NC, CO, or combinations of these to facilitate multiple (isolated) outputs. However, all these relays have single inputs to energize or de-energize the relay. In our discussions, only NO and NC relays are employed to describe the relay-based strategies.

Figures A.4 and A.5 illustrate the physical, logical (truth table), and timing relationships between the inputs and outputs for both NO and NC relays.

A relay with NO contact retains the open contact under no input. When it receives a voltage input, the coil gets energized and operates the output contacts to close to transmit the true signal to the next stage, working as a **buffer/repeater**.

A relay with NC contact retains the closed contact under no input. When it receives a voltage input, the coil gets energized and operates the output contacts to open or no signal transmission to the next stage, working as an **inverter**.

Figures A.6 and A.7 illustrate the realization of 2 input AND and 2 input OR circuits using a pair of NO relays. NOT circuits are realized by NC relays.

Using the same approach, it is possible to realize the gates for a variety of logical functions, such as NAND, NOR, XOR, XNOR. With these, we can derive any combinatorial logic for the implementation of a discrete automation strategy.

One more relay component that must be considered for the implementation of relay-based automation strategy is the **timer relay**. This relay, when energized, changes its output state after a programmed delay, as illustrated in Figure A.8.

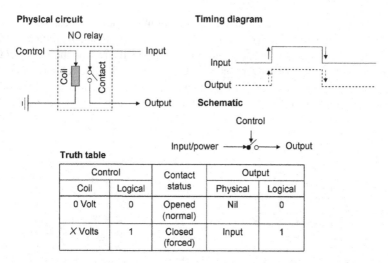

Figure A.4 Input/output relation in NO relay.

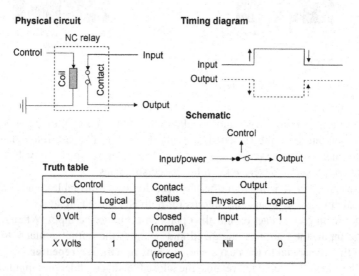

Figure A.5 Input/output relation in NC relay.

This is like any other relay but with a facility to program the delay time. The delay can be either on-delay or off-delay. Timer relays can be of either NO, NC, or CO configurations.

The typical wiring schematic for implementing the relay-based automation strategy using relay components is illustrated in Figure A.9. Here, a control signal moves from one stage to the other while satisfying the logic. The output from the previous stage becomes the control input for the next stage(s). In other words, implementation

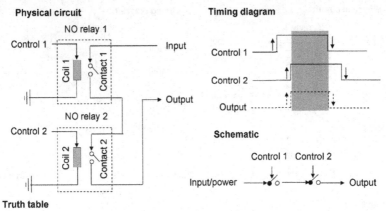

Truth table

Control 1		Control 2		Contact status		Output	
Coil 1	Logical	Coil 2	Logical	Contact 1	Contact 2	Physical	Logical
0 Volt	0	0 Volt	0	Opened (normal)	Opened (normal)	Nil	0
0 Volt	0	X Volts	1	Opened (normal)	Closed (forced)	Nil	0
X Volts	1	0 Volt	0	Closed (forced)	Opened (normal)	Nil	0
X Volts	1	X Volts	1	Closed (forced)	Closed (forced)	Input	1

Figure A.6 Realization of 2 input AND gate.

of relay logic means interconnecting various relays in a predefined and programmed manner.

A.2.1.1 Control Strategy Implementation

As already discussed, the relay-based automation strategy is applicable only for discrete process automation. Now that we've introduced the relay components for all the logic functions (repeater, inverter, NOT, AND, OR, timers, etc.), let's look into the implementation of relay-based automation strategies for the discrete automation cases discussed in Chapter 6.

Figure A.10 illustrates the water heating process as a discrete process with its instrumentation and human interface subsystems. The internal structure of the control subsystem will be discussed in the subsequent sections.

The functions and interconnections of the instrumentation devices and the human interface components with the control subsystem are explained in Table A.1.

As the control subsystem is built out of hardwired relay components, it is a hardware strategy. Before we start discussing various relay-based automation strategies, let's look into a few basic implementation schemes that are required. One such requirement is related to the issue of commands to the process.

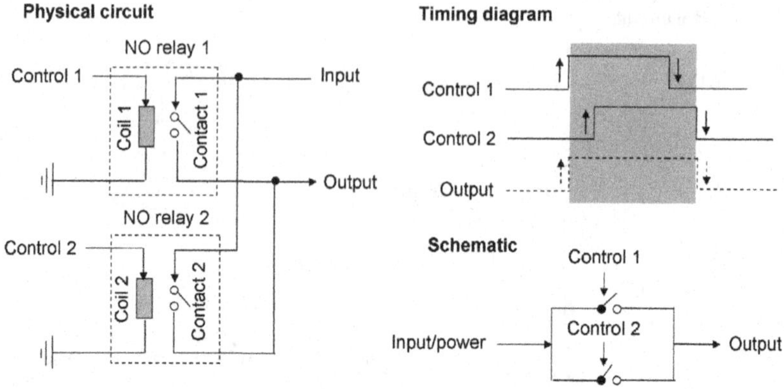

Truth table

Control 1		Control 2		Contact status		Output	
Coil 1	Logical	Coil 2	Logical	Contact 1	Contact 2	Physical	Logical
0 Volt	0	0 Volt	0	Opened (normal)	Opened (normal)	Nil	0
0 Volt	0	*X* Volts	1	Opened (normal)	Closed (forced)	Input	1
X Volts	1	0 Volt	0	Closed (forced)	Opened (normal)	Input	1
X Volts	1	*X* Volts	1	Closed (forced)	Closed (forced)	Input	1

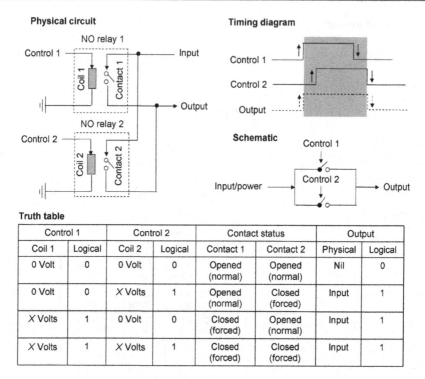

Figure A.7 Realization of 2 input OR gate.

Courtesy: www.gicindia.com

Figure A.8 Timer relay with delay time setting.

In most cases, the commands are generally of momentary type, meaning after the issue of the commands, the switch returns to its neutral position. This is required for meeting the following conditions:

- The switch position should not interfere with the functioning of the logic to issue commands automatically to the process. If the bi-stable command switch is employed, it remains in one of its two positions and blocks the issue of automatic commands by the logic.
- For safety reasons, if the bi-stable command switch remains in the closed position, it can issue the command to the process on powering the control subsystem.

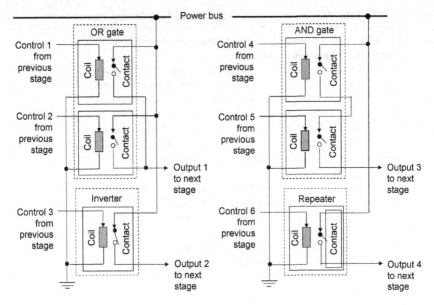

Figure A.9 Wiring scheme for relay strategy.

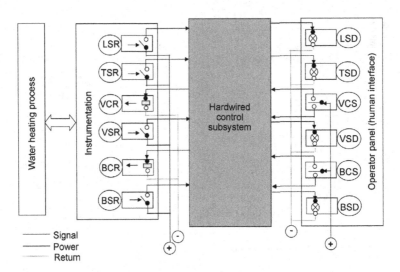

Figure A.10 Water heater with automation system (discrete).

For further processing of commands, the issued commands need to be remembered. Hence, there is a need to latch (or remember) the momentary commands as maintained commands, as illustrated in Figure A.11.

The scheme's timing diagram and its applications to open/close the valve and close/open the breaker are illustrated in Figures A.12 and A.13.

Table A.1 Instrumentation Devices and Their Interconnections

Between control subsystem and instrumentation (process)
Discrete instrumentation devices

1	LSR	Level status relay (preset value reached/not reached)
2	TSR	Temp status relay (preset value reached/not reached)
3	VCR	Valve control relay (command to open/close valve)
4	BCR	Breaker control relay (command to close/open breaker)
5	VSR	Valve supervision relay (opened/closed status of valve)
6	BSR	Breaker supervision relay (closed/opened status of breaker)

Between control subsystem and human interface (operator panel)
Discrete panel components

1	LSD	Level status indication lamp (on/off for reached/not reached)
2	TSD	Temp status indication lamp (on/off for reached/not reached)
3	VCS	Valve control switch (momentary commands to open/close)
4	BCS	Breaker control switch (momentary commands to close/open)
5	VSD	Valve status indication lamp (on/off for opened/closed)
6	BSD	Breaker status indication lamp (on/off for opened/closed)

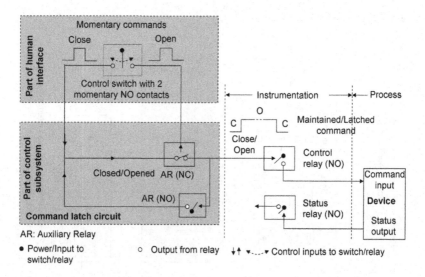

Figure A.11 Command latching scheme.

A.2.1.2 Open Loop Control—Discrete

Figure A.14 illustrates the implementation of simple discrete open loop control of level and temperature in the water heating process, as discussed in Chapter 6. Figure A.15 illustrates the associated timing in this strategy. Relay buffers/drivers are provided as a part of the control subsystem to isolate the control subsystem from the process and to repeat (reshaping and/or strengthening) the incoming signals in both directions. As

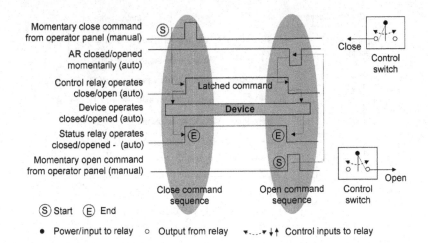

Figure A.12 Timing diagram of command latching.

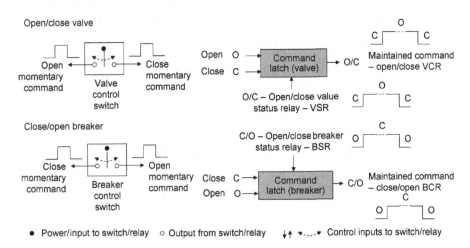

Figure A.13 Command latching for valve and breaker control.

seen here, there is absolutely no check on how long the valve is open or how long the breaker is closed, as this depends on the operator's manual interaction.

A.2.1.3 Sequential Control with Interlocks—Discrete

In sequential control with interlocks, there are no changes in either the instrumentation subsystem or in the human interface subsystem, as illustrated in Figure A.10. The change is only within the hardwired control subsystem (relay strategy implementation) to achieve sequential control with the interlock.

As illustrated in Figure A.16, the control subsystem built for sequential control with interlocks starts the heating process provided the interlock condition is satisfied, as discussed in Chapter 6. Figure A.17 illustrates the associated timing sequences.

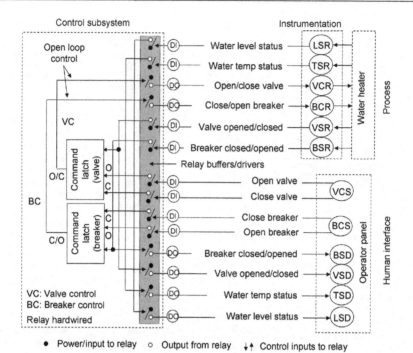

Figure A.14 Open loop control.

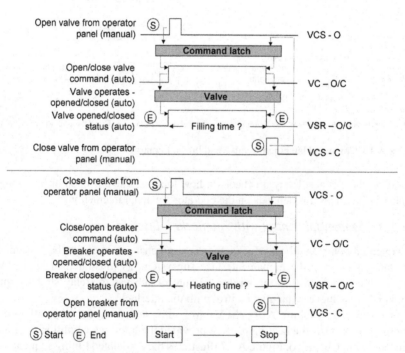

Figure A.15 Timing diagram of open loop control.

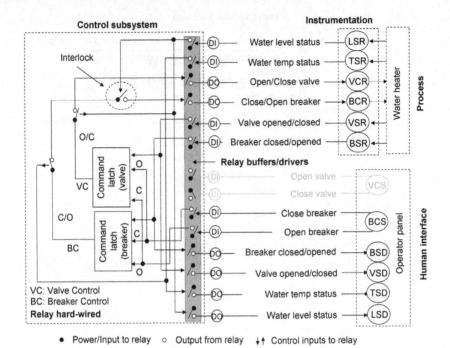

Figure A.16 Sequential controller with interlock.

A.2.1.4 Advantages and Disadvantages

The advantages of the relay-based strategy are as follows:

- Simple
- Easy to implement
- Easy to troubleshoot
- Does not require highly skilled persons for design, operation, and maintenance

The disadvantages are as follows:

- Occupies more space (bulky)
- Consumes more power
- Slow and sluggish response
- Unreliable (sensitive to environmental conditions, such as temperature, humidity, dust, vibration)
- Not maintenance-free (needs periodic preventive maintenance)
- Inflexible for modifications and extensions, as the scheme is hardwired

A.2.2 Solid State Technology

Some of the disadvantages of the relay-based strategy can be overcome by solid state technology. In this case, the electromechanical relay-based logic elements are

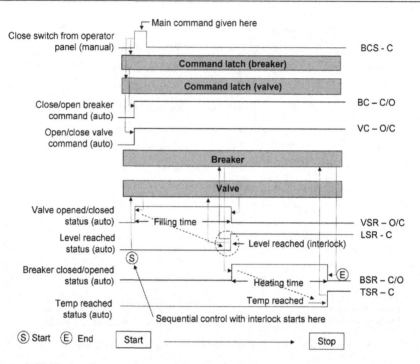

Figure A.17 Timing diagram of sequential control with interlocks.

replaced by their equivalent solid state elements on a one-to-one basis while maintaining all the technical features. Hence, the cases discussed previously for discrete process automation (open loop control and sequential control with interlocks) remain the same except for the change in control elements from relay-based to solid state, as shown in Figure A.18.

Using the basic logic functions, one can realize additional solid state logic functions, such as NAND, NOR, XOR.

A.2.2.1   Control Strategy Implementation

We devised a scheme in relay logic to convert momentary command into a maintained command (command latch), and Figure A.19 illustrates a similar scheme for solid state logic. Its timing diagram is shown in Figure A.20.

The following sections discuss the implementation of various control strategies with solid state logic elements in line with their relay counterparts.

A.2.2.2   Open Loop Control—Discrete

Figure A.21 illustrates the implementation of a solid state strategy for the discrete open loop control of the water heater process. All the aspects previously discussed for the relay-based strategy for open loop control are applicable here too.

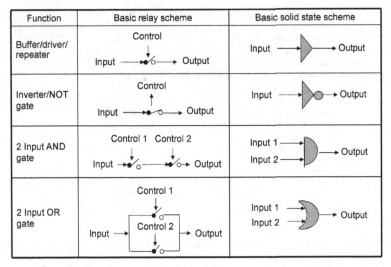

Figure A.18 solid state equivalents of relay logic elements.

Figure A.18 Solid state equivalents of relay logic elements.

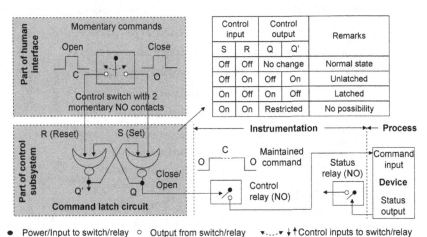

Figure A.19 Command latching scheme.

A.2.2.3   Sequential Control with Interlocks—Discrete

Figure A.22 illustrates the implementation of a solid state strategy for sequential control with interlocks of the water heater process. All the aspects previously discussed for relay-based strategy for open loop control are applicable here too.

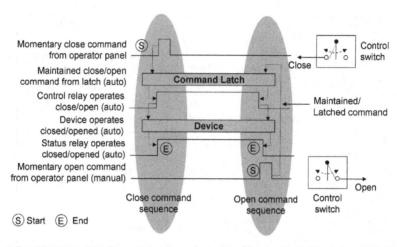

Figure A.20 Timing diagram for command latching.

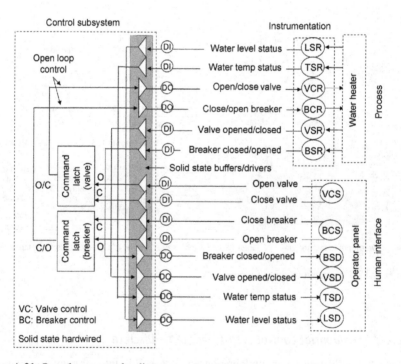

Figure A.21 Open loop control—discrete.

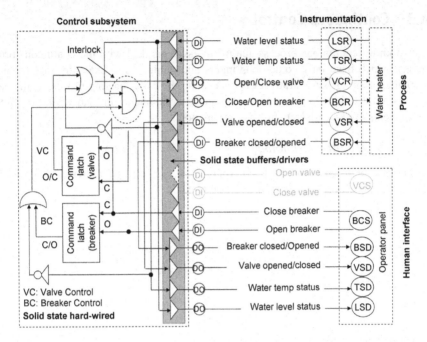

Figure A.22 Sequential control with interlocks.

In the previous illustrations, it is to be noted that the change is only in the hard-wired control subsystem (relay-based strategy), and there is no change in either the instrumentation or human interface subsystem.

A.2.2.4 Advantages and Disadvantages

The advantages of solid state technology are as follows:

- All the advantages of the relay-based strategy
- Compact (occupies less space)
- Consumes less power
- Faster response
- Maintenance-free (no need for periodic preventive maintenance)
- Reliable (not sensitive to environmental conditions, such as temperature, humidity, dust, vibrations)

The disadvantages are as follows:

- Inflexibility for modifications and extensions, as the scheme is hardwired (same as in relay-based strategy)

Early solid state logic elements were made of discrete components (circuits of resistors, capacitors, diodes, transistors, etc., mounted and wired on a board or a printed circuit board). Later, the discrete component-based elements moved to integrated circuits mounted on a printed circuit board.

A.3   Continuous Control

As we know, continuous process automation systems deal with the continuous pro-cesses, which have only continuous inputs and outputs. The example used for the implementation of these strategies is once again the water heating process with ana-log instrumentation devices and analog panel components instead of the discrete components used for discrete process automation.

A.3.1   Solid State Technology

In this technology, the most basic control element is a solid state **operational ampli-fier**, or simply an op amp, as shown in Figure A.23.

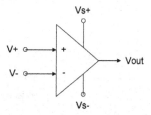

Figure A.23 Operational amplifier (op amp).

An op amp is basically a DC-coupled high gain electronic voltage amplifier with a differential input and, usually, a single-ended output. An op amp produces an out-put voltage that is typically much higher than the difference in input voltages of the order 10^6. The op amp, with some additional circuitry, can produce various control functions, as shown in Figure A.24.

Function	Symbol	Input-Output relation
Buffer/Repeater/ Voltage follower	Vin → ▷ → Vout	Vout = Vin
Summation amplifier	Vin1, Vin2 → [+ S +] → Vout	Vout = Vin1 + Vin2
Difference amplifier	Vin1, Vin2 → [+ D −] → Vout	Vout = Vin1 − Vin2
Comparator	Vin1, Vin2 → [+ C −] → Vout	Vout = High if Vin1 > Vin2, else Low

Figure A.24 Control elements of continuous process automation.

A.3.1.1   Control Strategy Implementation

The solid state components are applicable only to the implementation of automation strategies for a continuous process. We've discussed the solid state components for all the continuous functions (repeater/driver/buffer, summation, integration, differentiation, etc.), and now we'll look into the implementation of solid state automation strategies for the continuous process automation cases discussed in Chapter 6. Here, the control subsystem is built out of wired solid state components. Hence, this is also a hardware strategy.

A.3.1.2   Open Loop Control—Continuous

Figure A.25 illustrates the water heating system as a continuous open loop control with its instrumentation and human interface subsystems.

The details of the instrumentation devices and human interface panel components employed in the example are given in Table A.2.

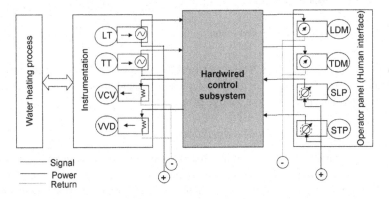

Figure A.25 Water heater automation (continuous).

Table A.2 Instrumentation Devices and Their Interconnections

Between control subsystem and instrumentation (process)
Analog instrumentation devices
1	LT	Water level transmitter
2	TT	Water temp transmitter
3	VCV	Variable control valve
4	VVD	Variable voltage drive

Between control subsystem and human interface (operator panel)
Analog panel components
1	LDM	Water level display meter
2	TDM	Water temp display meter
3	SLP	Desired water level setting potentiometer
4	STP	Desired water temp setting potentiometer

Figure A.26 illustrates the implementation of the open loop continuous control of a water heating process. As seen here, desired values for level and temperature are

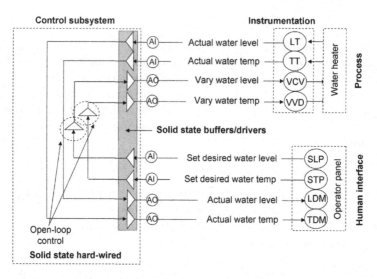

Figure A.26 Open loop control—continuous.

set independent of each other. The water heater produces the level and the temperature as per reference values. The outputs (actual level and actual temperature) stay at the set-point values until another change is made.

A.3.1.3 Closed Loop Control—Continuous

The implementation of a continuous closed loop control, as illustrated in Figure A.27, is true continuous control, as the output continuously follows the reference value. As seen in this scheme, the instrumentation and human interface requirements are similar to open loop control while the automation strategy is different.

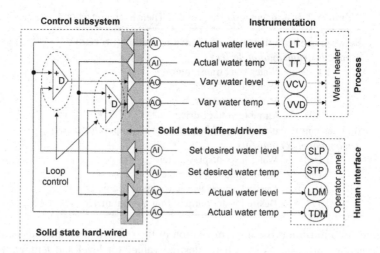

Figure A.27 Closed loop control—continuous.

A.4 Hybrid Control

Hybrid control deals with a combination of discrete and continuous processes that has both continuous and discrete I/O. The example used for the implementation of this strategy is once again the water heating process with input analog instrumentation devices, input and output analog panel components, and discrete output instrumentation devices.

A.4.1 Solid State Technology

In hybrid control, solid state technology must be used, as the process has both discrete and continuous signals. This is discussed in the following sections.

A.4.1.1 Control Strategy Implementation

We'll now revisit two hybrid strategies, two-step control and two-step control with dead-band, as discussed in Chapter 6. In these strategies, the control subsystems are built out of wired solid state components. Hence, these are also hardware strategies.

A.4.1.2 Two-Step Control

The two-step control is a combination of both discrete and continuous control and is a cost-effective approximation of continuous control. In fact, with two-step control, we can theoretically achieve true continuous control, but not practically because of hardware limitations of the final control elements.

We discussed two-step control strategy (with its two variants—with and without dead-band) and its advantages and disadvantages. Here, we'll discuss its implementation using solid state control components for both variants.

Figure A.28 illustrates the water heating process with its instrumentation and human interface subsystems. Contrary to open loop control, this closed loop control scheme employs both discrete and continuous instrumentation devices and continuous panel components in the human interface subsystem.

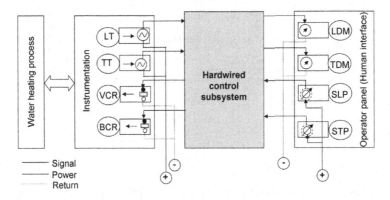

Figure A.28 Water heater with automation system (hybrid).

The list of analog and digital instrumentation devices and analog human interface panel components are given in Table A.3.

Figure A.29 illustrates the implementation of a simple two-step control strategy using solid state control components. The intention here is to maintain the water level and water temperature as per the set desired values. Here, the outputs (or results) follow the reference values in small discrete steps based on the accuracy/resolution of the instrumentation devices. The system is basically unstable as the control subsystem responds to even minor changes in the actual level and temperature.

Table A.3 Instrumentation Devices and Their Interconnections

Between control subsystem and instrumentation (process)
Analog instrumentation devices

1	LT	Water level transmitter
2	TT	Water temp transmitter

Discrete instrumentation devices

1	VCR	Valve control relay
2	BCR	Breaker control relay

Between control subsystem and human interface (operator panel)
Analog panel components

1	LDM	Water level display meter
2	TDM	Water temp display meter
3	SLP	Desired water level setting potentiometer
4	STP	Desired water temp setting potentiometer

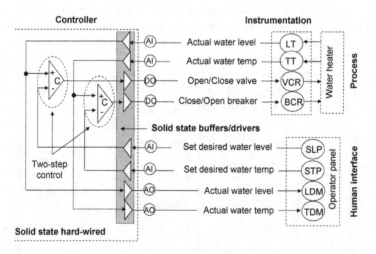

Figure A.29 Two-step control.

A.4.1.3 Two-Step Control with Dead-Band

As shown in Figure A.30, the implementation of the two-step control with dead-band requires some additions in the human interface subsystem (operator panel) to set the desired dead-bands for both level and temperature control while the rest is identical to simple two-step on/off control.

The revised list of the analog instrumentation and analog human interface panel components is given in Table A.4.

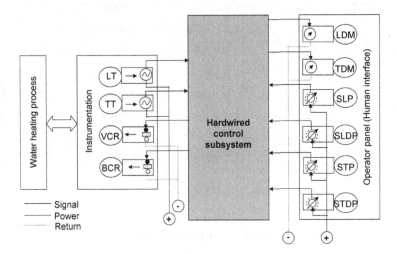

Figure A.30 Water heater with automation system (hybrid).

Table A.4 Instrumentation Devices and Their Interconnections

Between control subsystem and instrumentation (process)
Analog instrumentation devices

1	LT	Water level transmitter
2	TT	Water temp transmitter
3	VCR	Valve control relay
4	BCR	Breaker control relay

Between control subsystem interface (operator panel)
Analog panel components

1	LDM	Water level display meter
2	TDM	Water temp display meter
3	SLP	Desired water level setting potentiometer
4	SLDP	Desired water level dead-band setting potentiometer
5	STP	Desired water temp setting potentiometer
6	STDP	Desired water temp dead-band setting potentiometer

Figure A.31 illustrates the implementation of the two-step control strategy with dead-band using solid state components. As seen here, the higher the dead-band, the coarser the control while the lower the dead-band, the finer the control.

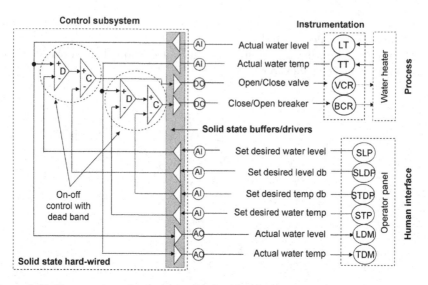

Figure A.31 Two-step control subsystem with dead-band.

A.5 FPGA- and FPAA-Based Controllers

As a natural extension of integrated circuit (IC) technology, one can go for advanced technologies, such as field programmable gate arrays (FPGAs), for realizing discrete process control subsystems and field programmable analog arrays (FPAAs) for continuous process control, or a combination of both FPGA and FPAA for two-step control. The programmability of FPGA- and FPAA-based control subsystems overcomes the inflexibility of the hardwired control subsystems toward modifications/extensions. These technologies fill the gap between the hardware-based and soft-wired (processor-based) technologies.

Appendix B
Processor

B.1   Introduction

The controller has four primary subsystems: power supply, processor, input/output, and communication. Out of these, the processor subsystem is the heart of the controller, as it controls and coordinates all the other subsystems as well as the operations of the controller. In this appendix, the construction and functions of the processor module are discussed in detail. We will discuss a hypothetical processor to explain the basic concepts involved.

B.2   Hardware structure

In this section, our example is an 8-bit hypothetical processor, as shown in Figure B.1. The processor discussed here is based on early architecture to explain the concepts for beginners in a more simple way.

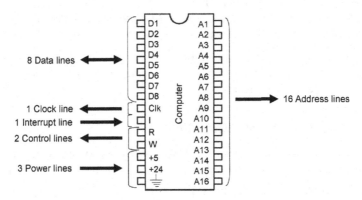

Figure B.1 Hypothetical processor.

B.2.1   Bus

The processor module and the functional modules are placed on the bus for communication, as illustrated in Figure B.2. In this example, the processor supports two

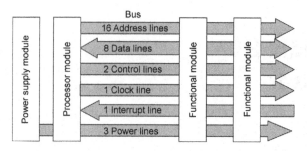

Figure B.2 Processor bus.

busses, internal and external. These busses have no difference logically as far as the communication between the processor and the functional modules is concerned. They differ only in their physical placement. The internal bus is an integrated part of the processor subsystem.

In this setup, the following lines are provided on the bus:

- Power lines for supplying the power to the processor and other functional modules
- Address lines for addressing the memory locations and the registers in the functional modules
- Data lines for carrying the data from memory/functional modules to the processor and vice versa
- Control lines for facilitating the read data from the memory/functional modules into the processor and the write data from the processor into the memory/functional modules
- Interrupt line for allowing the functional modules to interrupt the processor to draw its attention for some higher priority work
- Clock lines for providing the processor-generated clock signals for the functional modules for synchronization of their operations with the processor

In this example, only the important lines on the bus essential to understanding the basics of processor function are considered. In practice, there are many more lines on the bus for additional functions.

B.2.2 Address Space and Distribution

In our processor, 16 address lines are available which can address a maximum of 2^{16} or 65,536 locations with ranges as follows:

- Binary 0000 0000 0000 0000 to 1111 1111 1111 1111
- Octal 000000 to 177777
- Decimal 0 to 65,535

Figure B.3 illustrates the total address space available and its allocation in the processor.

The address space allocations are made as follows:

- First 61,440 addresses are for locations in the memory modules.
- Last 4,096 addresses are reserved for the registers in the functional modules.

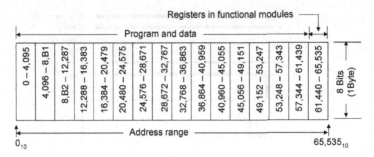

Figure B.3 Address space.

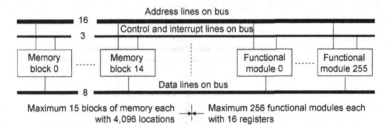

Figure B.4 Address space allocation.

Figure B.4 illustrates the allocation of the address space for addressing the memory locations and the registers in various functional modules.

With this arrangement, the processor can have a maximum of 15 memory blocks, each with up to 4,096 locations, and 256 functional modules, each with up to 16 registers.

B.2.3   Interfacing of Modules with Bus

The following subsections explain the interfacing of the functional modules of the controller with its bus.

B.2.3.1   Power Supply Module

Figure B.5 illustrates the interfacing of the power supply module with the bus. The power supply module receives the external power supply (230V AC or 48/24V DC), generates regulated +5V DC and +24V DC, and feeds this to the bus for processor and functional modules connected to the bus.

B.2.3.2   Processor Module

Figure B.6 illustrates the interfacing of the processor module with the bus. The processor module manages the address lines, data lines, and control lines for read

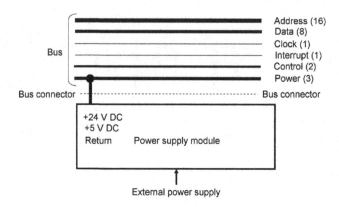

Figure B.5 Power supply module interfacing with bus.

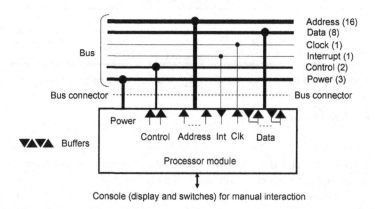

Figure B.6 Processor module interfacing with bus.

and write operations from/into the memory and registers in the functional modules. Additionally, the processor receives the interrupt signal (generated by other functional modules) over the bus and provides the clock signal on the bus for use by other functional modules.

B.2.3.3 Memory Module

Figure B.7 illustrates the interfacing of the memory module with the bus. The memory module uses the address lines, data lines, and control lines for placing the data on the bus from its location and writing the data from the bus into its location.

Figure B.8 illustrates the scheme for the application of the last four address lines (A12, A13, A14, and A15) for the selection of memory blocks. The other 12 address lines (A0–A11) use the facility built into the memory block for selection of the location within the block.

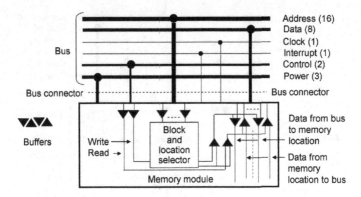

Figure B.7 Memory module interfacing with bus.

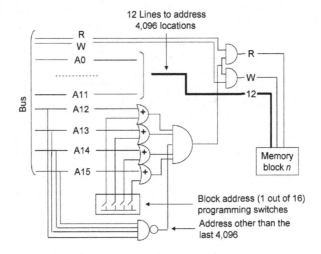

Figure B.8 Selection of memory block/location.

B.2.3.4 Functional Module

Figure B.9 shows the interfacing with the bus of the functional modules, such as the Input/Output (I/O) module, communication module, watchdog module. The selected functional module responds to the processor request over control lines either to place the data on the bus to be received by the processor or to receive the data from the bus placed by the processor. Further, the functional modules can send the interrupt signal provided they are allowed by the processor to interrupt.

Here, the selected functional module performs two operations on the bus: sends the data to the processor from one of its registers and receives the data from the processor for storage in one of its registers, as illustrated in Figure B.10.

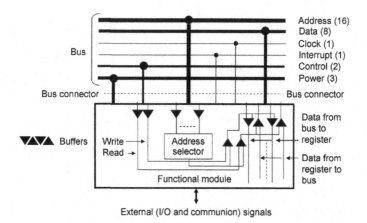

Figure B.9 Functional module interfacing with bus.

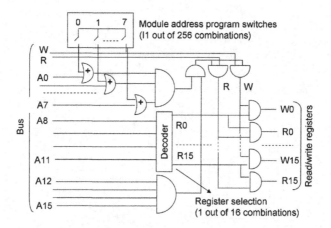

Figure B.10 Selection of registers in functional module.

B.2.3.5 Bus Extension (Parallel) Module

As illustrated in Figure B.11, the bus extension (parallel) module has no logical functions and it simply amplifies and drives the bus signals in both directions over the media. It is a physical link between two racks, extending the bus.

B.2.3.6 Bus Extension (Serial) Module

The bus extension (serial) module is identical to the bus extension (parallel) module, as far as bus interface is concerned.

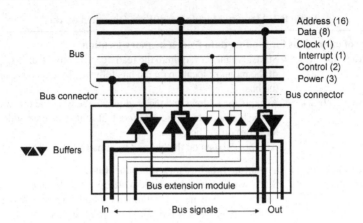

Figure B.11 Interfacing of bus extension module with bus.

B.2.3.7 Operations on the Bus

The processor performs two operations on the bus: receives the data from the memory/register in a functional module and sends the data to the memory/register in a functional module.

The processor is always the bus master, and all the other modules (memory, watchdog, I/O, communication, etc.) are like the slaves. Only the processor can read and write the data from the other modules over the bus. The operations over the internal and external bus are the same. The internal bus is an extension of the external bus, which reduces the communication load on the external bus and makes the communication of other functional modules faster. The following sections explain the read/write operation on the bus for memory locations as well as the registers in the functional modules.

B.2.3.8 Memory Module

Table B.1 illustrates the sequence of operations in reading from and writing to a memory location in a memory module.

B.2.3.9 I/O Module

Table B.2 illustrates the sequence of operations in reading from and writing to a register in the I/O module.

B.2.3.10 Communication Module

Table B.3 illustrates the sequence of operations in reading from and writing to a register in the communication module.

Table B.1 Memory Module Read/Write Operation on Bus

Sequence of operations for reading data from a memory location:

1	Processor	Starts, whenever required, the read operation by placing the address of the memory location on the bus followed by a read request on the bus.
2	Memory block	Accepts the address and the read request sent by the processor on the bus and places the data (content of the selected/addressed location) on the bus.
3	Processor	Accepts the data on the bus and terminates the read operation.

Sequence of operations for writing data into a memory location:

1	Processor	Starts, whenever required, the write operation by placing the address of the memory location on the bus. Then it places data to be written on the bus, followed by a write request.
2	Memory block	Accepts the address, the write request, and the data sent by the processor on the bus, then writes the data into its selected/addressed location.

Table B.2 I/O Module Read/Write Operations on the Bus

Sequence of operations for reading process inputs:

1	Process	Electronically maintains the latest values/states of its parameters at the input channels of the I/O module.
2	I/O module	Converts[a] the electronic inputs into data and transfers them to its data-in register. The data-in register always has the latest input data. This is done irrespective of whether the latest data is read by the processor or not.
3	Processor	Starts, whenever required, the read operation by placing the address of the data-in register of the I/O module on the bus followed by a read request.
4	I/O module	Accepts the address and the read request sent by the processor on the bus, and places the data (content of the selected data-in register) on the bus.
5	Processor	Accepts the data on the bus and terminates the read operation.

Sequence of operations for writing process outputs:

1	Processor	Starts, whenever required, the write operation by placing the address of the data-out register of the I/O module on the bus. It places data to be sent on the bus, followed by a write request.
2	I/O module	Accepts the address, write request, and the data sent by the processor on the bus, transfers them to its addressed data-out register, converts[b] the data into electronic signals, and drives the outputs channels.
3	Process	Accepts the electronic outputs received from the output channels of the functional module for further operations.

[a]Level in DI, analog to digital in AI, serial to parallel in PI
[b]Level in DO, digital to analog in AO, parallel to serial in PO

Table B.3 Communication Module Read/Write Operation on Bus

Sequence of operations for reading the message from the medium:

1	Medium	Feeds the electronic pulses serially into the communication I/F.
2	Communication	Collects the electronic pulses serially, converts them into bits, assembles them (serial to parallel conversion) into a message of 8 bits, and transfers them to its data-in register. The data-in register always has the latest message until it is overwritten by the next message. This is done irrespective of whether the latest message is read by the processor or not.
3	Processor	Starts, whenever required, the read operation by placing the address of the data-in register of the communication I/F on the bus followed by a read request on the bus.
4	Communication	Accepts the address and the read request sent by the processor on the bus and places the data (content of the data-in register) on the bus.
5	Processor	Accepts the data on the bus and terminates the read operation.

Sequence of operations for writing the message to the media:

1	Processor	Starts, whenever required, the write operation by placing the address of the data-out register of the communication I/F on the bus. It places data to be sent on the bus, followed by a write request.
2	Communication I/F	Accepts the address, the write request, and the data sent by the processor on the bus, transfers them to its data-out register, converts (parallel to serial) the data into electronic pulses, and sends them on the medium serially.
3	Medium	Accepts the electronic outputs received from the communication I/F for further transmission.

B.2.3.11 Watchdog Module

The watchdog works the same way as the other I/O modules.

Appendix C
Hardware-Software Interfacing

C.1   Introduction

In this appendix, the hardware and software interfacing of various functional modules is discussed. Without this interface, the software in the processor cannot access the functional modules for performing process data acquisition and process control. In other words, without the integration of hardware and software, no data transfer takes place between the processor and the functional modules.

C.2   Architectural Aspects

This section explains the architectural aspects of the processor, which are relevant for the hardware interfacing with the software.

C.2.1   Address Distribution

As seen in Figure C.1, the processor provides for 4,096 addresses (address range) to address the registers of the functional modules with the following distribution:

- Maximum 256 functional modules
- Maximum 16 registers per functional module

Generally, these addresses can be freely allotted to any functional module through the address selection facility within the functional module.

C.2.2   Processor Registers

We discussed the following about processor registers:

- Program instructions reside in the program memory area.
- Data resides either in the data memory area or in the registers of the functional modules, as the latter are identical to the memory locations.
- Execution of stored instructions in the program memory operates on the data in the data memory and/or the registers in the functional modules to produce results.

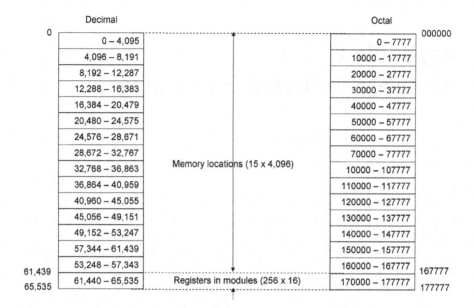

Figure C.1 Address distribution.

The processor is provided with the following registers:

- *Program counter:* A register to indicate the memory location address of the instruction under execution by the processor. On completion of the execution of every instruction, the program counter increments by either 1 or 3 automatically (see Section C.2.4) to indicate the address of the memory location of the next instruction in the program.
- *Accumulator:* A register to perform the following functions:
  - Hold the data read from memory/registers temporarily during its manipulation
  - Hold the result before transferring it to memory/registers
  - Act as one of the two operands in arithmetic/logical operations

C.2.3   Data Range in Memory/Registers

As illustrated in Figure C.2, the processor, an 8-bit machine, can accommodate a maximum of 8 bits (1 byte) of information content or supporting data that ranges as follows:

- Binary 00 000 000 to 11 111 111
- Octal 000 to 377
- Decimal 0 to 255

The content of a byte in the memory can be an instruction, an address, or data (arithmetic or logical) while the content in a register is always the data.

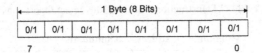

Figure C.2 Data range in memory/registers.

C.2.4   Instruction Formats

Here, basic instruction examples have been selected to address the concepts in the working of processor for automation strategy programming. In practice, the processors have an extensive repertoire of instructions.

For the program (instructions stored in the memory) to operate on the data in the memory/registers of the functional modules, we need to define the instructions for various operations for data movement, logical, arithmetic, and control operations. The following sections illustrate the various instruction formats.

The general format of the instruction has two fields, operation code, or opcode, and address. Further, some instructions do not need the address field while others need (2 bytes), as illustrated in Figure C.3.

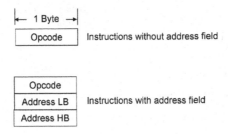

Figure C.3 Instruction formats.

The instructions without address fields occupy one location in the program memory and make the program counter increment by 1and on completion, to go to the next instruction in the program. Similarly, the instructions with address fields increase the program counter by 3, and on completion of their execution, to go to the next instruction in the program.

C.2.4.1   Data Movement Operations

Table C.1 illustrates the data movement instructions and their operations.

C.2.4.2   Logical Operations

Table C.2 illustrates the logical instructions and their operations.

Table C.1 Instructions for Data Movement Operations

Opcode	Mnemonic	Address	Comments
000_8 (00 000 000)	CLA	–	Clear accumulator
001_8 (00 000 011)	LDA	X	Load content of location X in accumulator
002_8 (00 000 100)	STA	X	Store content of accumulator in location X
003_8–007_8	Reserved		

Table C.2 Instructions for Logical Operations

Instruction	Mnemonic	Address	Comments
010_8 (00 001 000)	ANA	X	Logically AND location X content with accumulator content
011_8 (00 001 001)	ORA	X	Logically OR location X content with accumulator content
012_8 (00 001 010)	CMA	–	Logically complement content of accumulator
013_8 (00 001 111)	SRA	–	Logically shift right by 1 bit of content of accumulator
014^8 (00 001 100)	SLA	–	Logically shift left by 1 bit of content of accumulator
015_8–017_8	Reserved		

C.2.4.3 Arithmetic Operations

Table C.3 illustrates the arithmetic instructions and their operations.

Table C.3 Instructions for Arithmetic Operations

Instruction	Mnemonic	Address	Comments
020_8 (00 010 000)	ADA	X	Add content of location X to accumulator content
021_8 (00 010 001)	SBA	X	Subtract content of location X from accumulator content
022_8 (00 010 010)	INA	–	Increment content of accumulator by 1
023_8 (00 010 011)	DCA	–	Decrement content of accumulator by 1
024_8–027_8	Reserved		

C.2.4.4 Control Operations

Table C.4 illustrates the control instructions and their operations.

Table C.4 Instructions for Control Operations

Instruction	Mnemonic	Address	Comments
030_8 (00 011 000)	JMP	X	Jump to location X
031_8 (00 011 001)	JNZ	X	If content of accumulator is nonzero, jump to location X
032_8 (00 011 010)	JMZ	X	If content of accumulator is zero, jump to location X
033_8 (00 011 011)	CSR	X	Call subroutine after saving the calling program status in stack
034_8 (00 011 011)	RSR	X	Return from subroutine after restoring the called program status from stack
035_8 (00 011 100)	HLT	–	Halt program execution
$036_8, 037_8$	Reserved		

C.2.5 Program Interfacing with Functional Modules

Each functional module is provided with a few addressable registers for interfacing with the program:

- Status registers to hold the information related to the functioning and control of the module and its operation
- Data registers to hold the data received from the process and the data to be sent to the process

The general formats of both status and data registers are illustrated in Figure C.4.

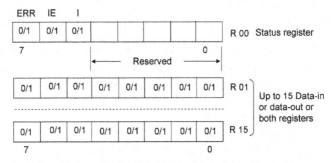

ERR: Error bit (0: module healthy and 1: module faulty) - set by module
IE: Interrupt enable bit (0: disable and 1: enable) - set by processor
I: Interrupt bit (0: no interrupt and 1: interrupt) - set by module

Figure C.4 Format for status and data registers in functional module.

C.2.5.1 Status Register

The explanations of bits 7, 6, and 5 in the status register are as follows:

- Bit 7, or ERR bit, indicates the health of the functional module—if 0, the module is healthy, and if 1, the module is not healthy (or faulty). This bit is set internally by the module and is read only by the processor. The module clears this bit (or sets it to 0) internally once the fault is cleared or the module is healthy. The processor has no control over this bit except that it can read it.

- Bit 6, or IE bit, indicates whether the functional module is enabled to interrupt the processor or not—if 0, the module is not enabled, and if 1, the module is enabled. The processor controls this bit (setting it to 0 or to 1), and the functional module has no control over this bit.
- Bit 6, or I bit, indicates whether an interrupt in the module is pending or not—if 0, no interrupt is pending, and if 1, an interrupt is pending. The interrupt is generated by the functional module (setting it to 1), and the module clears the bit internally once the processor reads it (resetting it to 0). This is also a read-only bit by the processor. The bit is controlled by the module, and the processor has no control over it.

C.2.5.2   Data Registers

The data registers are of two types:

- Data-in registers to hold the data received from the process. The functional module controls the contents in this register, and the processor has no control over it.
- Data-out registers to hold the data sent to the process. The processor controls the contents in this register, and the functional module has no control over it.

The following sections illustrate the specific formats of status and data registers for different types of functional modules.

C.2.6   Interfacing of Functional Modules with Software

The following sections explain the status and the data registers in I/O modules for interfacing them with the software.

C.2.6.1   Digital Input

Figure C.5 illustrates the formats of status and data registers in a DI module. This module supports eight inputs and hence requires two registers—one for status (R00) and one for data-in (R01). While the explanations for bits ERR and IE are the same as explained earlier, bit I gets set provided bit IE is set by the processor and one of the following conditions take place:

- Module is or becomes faulty (bit ERR gets set)
- Arrival of new data in one or more channel (bit ND gets set)

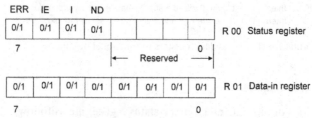

Figure C.5 Status and data registers in DI module.

C.2.6.2 Digital Output

Figure C.6 illustrates the formats of status and data registers in a DO module. This module supports eight outputs and hence requires two registers—one for status (R00) and one for data-out (R01). While the explanations for bits ERR and IE remain the same, bit I gets set provided bit IE is set by the processor and the module is or becomes faulty (bit ERR gets set).

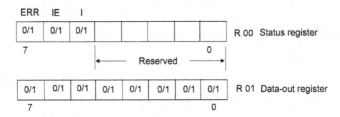

Figure C.6 Status and data registers in DO module.

C.2.6.3 Analog Input

Figure C.7 illustrates the formats of status and data registers in an AI module. This module supports four inputs and hence requires five registers—one for status (R00) and four for data-input (R01–R4). While the explanations for bits ERR and IE remain the same as explained, bit I gets set provided bit IE is set by the processor and the module is or becomes faulty (bit ERR gets set).

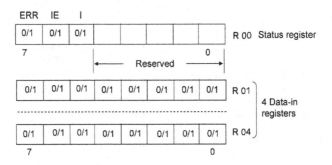

Figure C.7 Status and data registers in AI module.

C.2.6.4 Analog Output

Figure C.8 illustrates the formats of status and data registers in an AO module. This module supports two outputs and hence requires three registers—one for status (R00) and two for data-output (R01 and R02). While the explanations for bits ERR

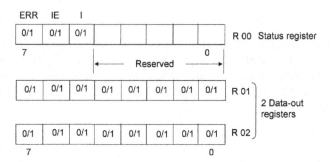

Figure C.8 Status and data registers in AO module.

and IE remain the same, bit I gets set provided bit IE is set by the processor and the module is or becomes faulty (bit ERR gets set).

C.2.6.5   Pulse Input

Figure C.9 illustrates the formats of status and data registers in a PI module. This module supports four inputs and hence requires five registers—one for status (R00) and four for data-input (R01–R4). While the explanations for bits ERR and IE remain the same, bit I gets set provided bit IE is set by the processor and the module is or becomes faulty (bit ERR gets set) or any counter becoming full (IBFn gets set).

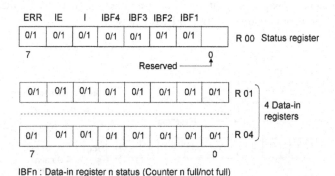

IBFn : Data-in register n status (Counter n full/not full)

Figure C.9 Status and data registers in PI module.

C.2.6.6   Pulse Output

Figure C.10 illustrates the formats of status and data registers in a PO module. This module supports four inputs and hence requires five registers—one for status (R00) and four for data-input (R01–R4). While the explanations for bits ERR and IE remain the same as explained, bit I gets set provided bit IE is set by the processor and the module is or becomes faulty (bit ERR gets set) or the counter becoming empty (IBEn gets set).

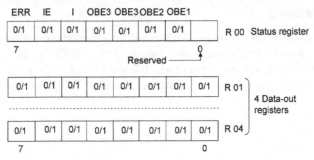

OBEn: Data-out register n status (counter n empty or not empty)

Figure C.10 Status and data registers in PO module.

C.2.6.7 Communication

Figure C.11 illustrates the formats of status and data registers in a communication module. This module supports one bidirectional communication channel and hence requires three registers—one for status (R00), one for data-in (R01), and one for data-out (R02). While the explanations for bits ERR and IE remain the same, bit I gets set provided bit IE is set by the processor and one or more of the following conditions are met:

- The module is or becomes faulty (bit ERR gets set).
- Send (data-out) register becomes empty with the completion of parallel to serial conversion or after pushing out the 8-bit parallel data serially to the communication medium (bit OBE gets set)
- Receive (data-in) register becomes full with the completion of serial to parallel conversion or after assembling the serial data received from the communication medium into 8-bit parallel data in the data-in register (bit IBF gets set)

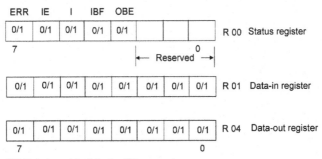

IBF: Data-in register full after S/P conversion

OBE: Data-out register empty after P/S conversion

Figure C.11 Status and data registers in communication module.

C.2.6.8 Watchdog

Figure C.12 illustrates the formats of status and data registers in a watchdog module. This module also requires two registers—one for status (R00) and the other for data-out (R01). While the explanations for bits ERR and IE remain the same, bit I gets set provided bit IE is set by the processor and the module is or becomes faulty (bit ERR gets set).

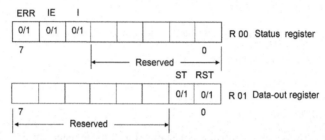

RST: Reset re-trigger mono-stable multi-vibrator reset (write only by processor))
ST: Announce nonfatal fault (write only by processor)

Figure C.12 Status and data registers in watchdog module.

Appendix D
Instruction Set of Processor

D.1 Introduction

The following information provides a summary of the basic programming instructions and their functions in the processor.

D.2 Data Movement Operations

Opcode	Mnemonic	Address	Comments
000_8 (00 000 000)	CLA	–	Clear accumulator content
001_8 (00 000 011)	LDA	X	Load content of location X in accumulator
002_8 (00 000 100)	STA	X	Store content of accumulator in location X
003_8–007_8	Reserved		

D.3 Logical Operations

Instruction	Mnemonic	Address	Comments
010_8 (00 001 000)	ANA	X	Logically AND content of location X with content of accumulator
011_8 (00 001 001)	ORA	X	Logically OR content of location X with content of accumulator
012_8 (00 001 010)	CMA	–	Logically complement content of accumulator
013_8 (00 001 111)	SRA	–	Logically shift right content of accumulator by 1 bit
014_8 (00 001 100)	SLA	–	Logically shift left content of accumulator 1 bit
015_8–017_8	Reserved		

D.4 Arithmetic Operations

Instruction	Mnemonic	Address	Comments
020_8 (00 010 000)	ADA	X	Add content of location X to content of accumulator
021_8 (00 010 001)	SBA	X	Subtract content of location X from content of accumulator
022_8 (00 010 010)	INA	–	Increase content of accumulator by 1
023_8 (00 010 011)	DCA	–	Decrease content of accumulator by 1
024_8–027_8	Reserved		

D.5 Control Operations

Instruction	Mnemonic	Address	Comments
030_8 (00 011 000)	JMP	X	Jump to location X
031_8 (00 011 001)	JNZ	X	Jump to location X, if content of accumulator is zero
032_8 (00 011 010)	JMZ	X	Jump to location X, if content of accumulator is nonzero
033_8 (00 011 011)	CSR	X	Call subroutine[a] at location X
034_8 (00 011 011)	RTS	–	Return from subroutine[b]
035_8 (00 011 100)	HLT	–	Halt program execution
036_8, 037_8	Reserved		

[a]Hardware stores the program status (PC and AC) in stack (first-in/last-out) and branches to location X to service the subroutine.
[b]Hardware restores the called program status (PC and AC) from the stack (first-in/last-out) and returns to the called program.

These are the instructions provided in the hypothetical processor considered in the book. The readers can try out the programming examples in the book with the instruction set available in any commercial computer.

Appendix E
Basics of Programming

E.1   Introduction

This appendix gives some insight into the basics of programming and its methods in lower levels (machine and assembly level).

E.2   Lower-Level Programming

The lower-level programming methods use machine language and assembly language, as discussed in the subsequent sections.

E.2.1   Machine Level

The starting point of programming is at machine level. This calls for the full knowledge of the processor architecture and its instruction set. Primarily, the coding in machine level is done with combinations of 0s and 1s. The loading of the program and the data is done manually. This method, which is not suitable for large programs, is used for inputting small programs for booting the machine and to hand over control to an interactive terminal. To facilitate this, computers provided an interactive console (hardwired) with switches for inputting the combinations of 0s and 1s into the addressed memory locations and console lamps to display the contents of the addressed locations. This approach has been subsequently replaced by a booting program (to start the computer on powering on) that is resident in the nonvolatile memory of the computer.

E.2.2   Assembly Level

Because it is very inconvenient and causes errors to work with 0s and 1s for coding, assembly-level programming replaced machine-level programming. Assembly-level programming also calls for the full knowledge of the processor architecture and its instruction set. The only difference is that mnemonics are used to abbreviate various instructions, addresses, labels, and data while coding the program. The **assembler**, a program supplied by the controller vendor, converts the assembly-level program into its machine equivalent executable code. Here, each line in the assembly-level program gets converted into one machine-level instruction (one-to-one). Assembler programs are very effective in developing system programs, as they can exploit the architecture of the processor.

E.3 Programming Examples

In the following examples, it is assumed that the first 256 locations (addresses from 000000_8 to 000377_8) in the memory are allocated for the user program (instructions) and the next 256 locations (addresses from 001000_8 to 002777_8) are reserved for user data. Further, the program starts from location 0 (000000_8) automatically on power-on of the processor.

When the processor wants to read from or write to memory locations or the registers in the functional modules for execution of an instruction, the address becomes part of the instruction. As the address is 16 bits, it is stored along with the instructions in 2 bytes (lower byte followed by higher byte) consecutively, as seen in the following programming examples. The programs illustrate both the machine level and its assembly language equivalent.

The following procedures are adopted in the programming examples discussed in this appendix:

- An image of the input data read from the registers in the input modules is created in input image buffer (IPIB) in the memory (data acquisition).
- The program operates on the latest available data in the input image buffer (IIPIB) and stores the result in output image buffer (OPIB) in the memory.
- The contents of OPIB are written into the registers of the output modules for effecting the process control.

With these arrangements, the registers in the Input/Output (I/O) modules are de-linked from the program. *In other words, the program always interfaces with the input image and output image buffers.* Following are the programming examples that employ this approach to handle I/O for process data acquisition, data analysis/decision, process control, and communication.

Further, in all the programs, before the input and output scans, the health of the I/O modules is checked and the watchdog is activated if the modules are found faulty. The schematic and the address allocation for the watchdog module are given in Figure E.1. This is common to all the subsequent programming examples.

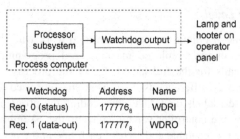

Watchdog	Address	Name
Reg. 0 (status)	177776_8	WDRI
Reg. 1 (data-out)	177777_8	WDRO

Figure E.1 Watchdog module—schematic and address allocation.

E.3.1 Programming with Digital I/O

Problem: To acquire discrete inputs from the process and display them on indication lamps on the operator panel

This is an example of reading the states of discrete inputs from the process using a DI module and driving the indication lamps on the operator panel using a DO module. Figure E.2 illustrates the schematic and allocation of the address to DI and DO modules.

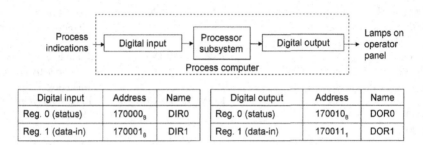

Digital input	Address	Name	Digital output	Address	Name
Reg. 0 (status)	170000_8	DIR0	Reg. 0 (status)	170010_8	DOR0
Reg. 1 (data-in)	170001_8	DIR1	Reg. 1 (data-in)	170011_1	DOR1

Figure E.2 Schematic and address allocation.

Figure E.3 illustrates the program flow chart.

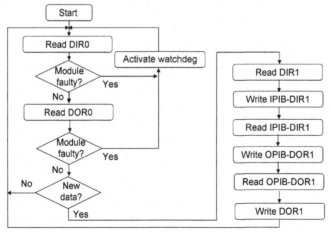

IPIB, OPIB: Input image and output image buffers in memory

Figure E.3 Flow chart—digital I/O program.

Table E.1 illustrates the program in assembly level with its machine equivalent machine code.

Table E.1 Machine/Assembly-Level Program for Digital I/O

Memory Location Address[a]	Content	Label	Instruction		Comments
Program area					
000000	001	STR	LDA	DIR0	Read status register
000001	000				DIR0
000002	170				
000003	010		ANA	XXX	Extract bit 7
000004	400				(diagnostic bit) from
000005	000				DIR0
000006	031		JNZ	WDG	If module is faulty,
000007	060				jump to activate
000010	000				watchdog
000011	001		LDA	DOR0	Read status register
000012	010				DOR0
000013	360				
000014	010		ANA	XXX	Extract bit 7
000015	000				(diagnostic bit) from
000016	001				DOR0
000017	031		JNZ	WDG	If module is faulty,
000020	060				jump to activate
000021	000				watchdog
000022	001		LDA	DIR0	Read status register
000023	000				DIR0
000024	360				
000025	010		ANA	YYY	Check for new data
000026	201				
000027	001				
000030	032		JMZ	STR	Go to start if no new
000031	000				data
000032	000				
000033	001		LDA	DIR1	Read data-in register
000034	000				DIR1
000035	361				
000036	002		STA	IPIB-DIR1	Store in input image
000047	100				buffer
000040	001				
000041	001		LDA	IPIB-DIR1	Read from input
000042	101				image buffer
000043	001				
000044	002		STA	OPIB-DOR1	Store in output image
000045	100				buffer
000046	001				
000047	001		LDA	OPIB-DOR1	Read from output
000050	100				image buffer
000051	001				

Table E.1 Machine/Assembly-Level Program for Digital I/O (Continued)

Memory Location Address[a]	Content	Label	Instruction		Comments
			Assembly-Level Instruction		
000052	002		STA	DOR1	Store in data-out
000053	011				register DOR1
000054	360				
000055	030		JMP	STR	Repeat program
000056	000				
000057	000				
000060	001	WDG	LDA	ZZZ	Load mask for
000061	002				extraction of
000062	000				watchdog bit
000063	002		STA	WDR1	Drive watchdog
000064	111				
000065	111				
000066	030		JMP	STR	Repeat program
000067	000				
000070	000				
Data area					
000400	200	XXX			For extraction of diagnostic bit
000401	010	YYY			For extraction of new data bit
000402	002	ZZZ			For driving watchdog bit
000500		IPIB-DIR1			
000501		OPIB-DOR1			

[a]Address represents program counter in program area.

E.3.2 Programming with Analog I/O

Problem: To acquire one continuous input from the process and display it on a meter on the operator panel

This is an example of reading the value of a continuous parameter from the process using an AI module (channel 0) and driving in onto the display meter on the operator panel using an AO module (channel 0). Figure E.4 illustrates this schematic, Figure E.5 the flow chart, and Table E.2 shows its program in assembly-level languages.

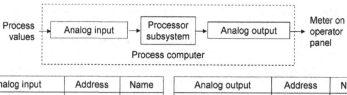

Analog input	Address	Name
Reg. 0 (Status)	170020_8	AIR0
Reg. 1 (Data-in 0)	170021_8	AIR1
Reg. 2 (Data-in 1)	$1700E_8$	AIR2
Reg. 3 (Data-in 2)	170023_8	AIR3
Reg. 4 (Data-in 4)	170024_8	AIR4

Analog output	Address	Name
Reg. 0 (Status)	170030_8	AOR0
Reg. 1 (Data-out 0)	170031_8	AOR1
Reg. 2 (Data-out 1)	170032_8	AOR2

Figure E.4 Schematic and address allocations.

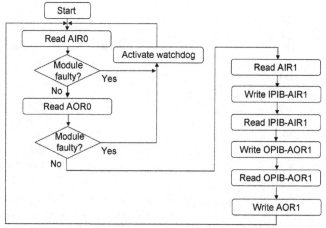

IPIB, OPIB: Input image and output image buffers in memory

Figure E.5 Flow chart for analog I/O program.

Table E.2 Assembly-Level Program for Analog I/O

Label	Instruction		Comments
Program area			
STR	LDA	AIR0	Load status register AIR0
	ANA	XXX	Extract bit 7 (diagnostic bit)
	JNZ	WDG	If module is faulty, jump to activate watchdog
	LDA	AOR0	Load status register AOR0
	ANA	XXX	Extract bit 7 (diagnostic bit)
	JNZ	WDG	If module is faulty, jump to activate watchdog
	LDA	AIR1	Load data-in register AIR1

Table E.2 Assembly-Level Program for Analog I/O (Continued)

Label	Instruction		Comments
	STA	IPIB-AIR1	Store in input image buffer IPIB-AIR1
	LDA	IPIB-AIR1	Read input image buffer IPIB-AIR1
	STA	OPIB-AOR1	Store in output image buffer OPIB-AOR1
	LDA	OPIB-AOR1	Load output image buffer OPIB-AOR1
	STA	AOR1	Store in data-out register AOR1
	JMP	STR	Repeat program
WDG	LDA	YYY	Load mask for extraction of watchdog bit
	STA	WDR1	Drive watchdog
	JMP	STR	Repeat program
Data area			
XXX	'200'		Mask for extraction of diagnostic bit
YYY	'002'		Mask for driving watchdog bit
IPIB-AIR1			Input image
OPIB-AOR1			Output image

E.3.3   Programming with Pulse I/O

Problem: To acquire one pulse input from the process and display it on a counter on the operator panel

This is an example of reading the pulses from the process using a PI module (channel 0) and driving them onto the display counter on the operator panel using a PO module (channel 0). Figure E.6 illustrates the schematic, Figure E.7 the flow chart, and Table E.3 shows its program in assembly-level language.

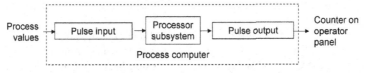

Pulse input	Address	Name	Pulse output	Address	Name
Reg. 0 (Status)	170040_8	PIR0	Reg. 0 (Status)	170050_8	POR0
Reg. 1 (Data-in 0)	170041_8	PIR1	Reg. 1 (Data-out 0)	170051_8	POR1
Reg. 2 (Data-in 1)	170042_8	PIR2	Reg. 2 (Data-out 1)	170052_8	POR2
Reg. 3 (Data-in 2)	170043_8	PIR3	Reg. 3 (Data-out 2)	170052_8	POR3
Reg. 4 (Data-in 4)	170044_8	PIR4	Reg. 4 (Data-out 3)	170052_8	POR4

Figure E.6 Schematic and address allocation.

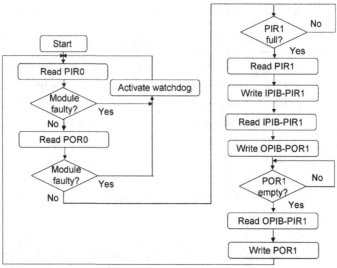

IPIB, OPIB: Input image and output image buffers in memory

Figure E.7 Flow chart for pulse I/O program.

E.3.4 Programming with Communication

Problem: To receive the message from Medium 1 and to send it to Medium 2

This is an example of reading the incoming message from medium 1 using communication module 1 and sending it to medium 2 over communication module 2. Figure E.8 illustrates the schematic, Figure E.9 the flow chart, and Table E.4 shows its program in machine- and assembly-level languages.

Table E.3 Assembly-Level Program for Pulse I/O

Label	Instruction		Comments
Program area			
STR	LDA	PIR0	Load status register PIR0
	ANA	XXX	Extract bit 7 (diagnostic bit)
	JNZ	WDG	If module faulty, jump to activate watchdog
	LDA	POR0	Load status register PIR0
	ANA	XXX	Extract bit 7 (diagnostic bit)
	JNZ	WDG	If module faulty, jump to activate watchdog
LOOP1	LDA	PIR0	Load status register PIR0
	AND	YYY1	Check for counter full

Table E.3 Assembly-Level Program for Pulse I/O (Continued)

Label	Instruction		Comments
	JZ	LOOP1	If counter not full, wait
	LDA	PIR1	Load data-in register PIR1
	STA	IPIB-PIR1	Store in input image buffer IPIB-PIR1
	LDA	IPIB-PIR1	Load input image buffer IPIB-PIR1
	STA	OPIB-POR1	Store in output image buffer OPIB-POR1
LOOP2	LDA	POR0	Load status register POR0
	ANA	YYY2	Check for counter empty
	JZ	LOOP2	If counter not empty, wait
	LDA	OPIB-POR1	Load output image buffer OPIB-POR1
	STA	POR1	Store in data-out register POR1
	JMP	STR	Repeat program
WDG	LDA	ZZZ	Load mask for extraction of watchdog bit
	STA	WDR1	Drive watchdog
	JMP	STR	Repeat program
Data area			
XXX	'200'		Mask for extraction of diagnostic bit
YYY1	'002'		Mask for counter full
YYY2	'002'		Mask for counter empty
ZZZ	'002'		Mask for driving watchdog bit
IPIB-PIR1			Input image
OPIB-POR1			Output image

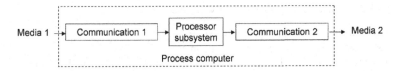

Communication 1	Address	Name		Communication	Address	Name
Reg. 0 (Status)	170060_8	CMR0-1		Reg. 0 (Status)	170070_8	CMR0-2
Reg. 1 (Data-in)	170061_8	CMR1-1		Reg. 1 (Data-in)	170071_8	CMR1-2
Reg. 2 (Data-out)	170062_8	CMR2-1		Reg. 2 (Data-out)	170072_8	CMR2-2

Figure E.8 Schematic and address allocations.

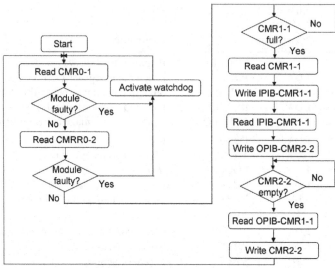

IPIB, OPIB: Input image and output image buffers in memory

Figure E.9 Flow chart for communication program.

E.3.5 *Programming with Interrupt*

Problem: To service an alarm from the process *immediately* on its occurrence and to drive the audiovisual alarm

This is similar to the first example in which the process indications are acquired using a digital input module and sent to lamps on a mimic panel using a digital output module—this occurs through continuous scanning for new data. The only difference is that the hardware interrupt facility is employed to perform the job immediately provided *the digital input module is enabled to interrupt the processor.* In programming with interrupt, the following occurs:

- Processor enables the digital input module to interrupt.
- Processor keeps executing the current program.
- Digital input module, on change of state in any of its inputs, sets new data (ND) bit in its status register and interrupts the processor.
- Processor, after completion of the execution of the current instruction, checks for pending interrupts before moving on to the execution of the next instruction in the current program.
- Processor, on recognition of an interrupt, suspends further execution of the current program.
- Processor automatically saves the latest execution status of the current (interrupted) program (contents of program counter, accumulator) in the first-in/last-out stack in the memory.

Table E.4 Assembly-Level Program for Communication

Label	Instruction		Comments
Program area			
STR	LDA	CMR0-1	Load status register CMR0-1
	ANA	XXX	Extract bit 7 (diagnostic bit)
	JNZ	WDG	If module faulty, jump to activate watchdog
	LDA	CMR0-2	Load status register CMR0-2
	ANA	XXX	Extract bit 7 (diagnostic bit)
	JNZ	WDG	If module faulty, jump to activate watchdog
LOOP1	LDA	CMR0-1	Load status register CMR0-1
	AND	YYY1	Check for buffer full
	JZ	LOOP1	If message not full, wait
	LDA	CMR1-1	Load data-in register CMR1-1
	STA	IPIB-CMR1-1	Store in input image buffer IPIB-CMR1-1
	LDA	IPIB-CMR1-1	Load input image buffer IPIB-CMR1-1
	STA	OPIB-CMR2-2	Store in output image buffer OPIB-CMR2-2
LOOP2	LDA	CMR0-2	Load status register CMRO-2
	ANA	YYY2	Check for buffer empty
	JZ	LOOP2	If counter not empty, wait
	LDA	OPIB-CMR2-2	Load output image buffer OPIB-CMR2-2
	STA	CMR2-2	Store in data out register CMR2-2
	JMP	STR	Repeat program
WDG	LDA	ZZZ	Load mask for extraction of watchdog bit
	STA	WDR1	Drive watchdog
	JMP	STR	Repeat program
Data area			
XXX	'200'		Mask for extraction of diagnostic bit
YYY1	'010'		Mask for buffer full
YYY2	'020'		Mask for buffer empty
ZZZ	'002'		Mask for driving watchdog bit
IPIB-CMR1-1			Input image
OPIB-CMR2-2			Output image

- Processor branches to general interrupt service routine to identify the source of the interrupt (digital input module in the present case).
- Processor branches to digital input service routine.

- Processor reads the data-in register of the digital input module and stores it in the input image buffer in the memory.
- Processor returns to the interrupted program by restoring the status of the interrupted program from the first-in/last-out stack in the memory (resumption of the interrupted program).

Figure E.10 and Table E.5 illustrate the flow chart and the program.

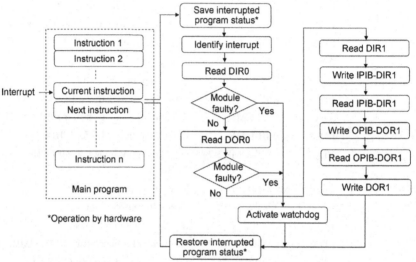

Figure E.10 Flow chart for interrupt program.

It is also possible to use the feature of module interrupt when there are faults (by setting its diagnostic bit) to prevent module faults from being checked in the previous program.

E.3.6 Assembling of Program

The automation programs written (coded) in assembly-level language need be converted into their machine executable equivalents for execution by the processor. This process is called **assembling** and is carried out by a special program called **assembler**, which is supplied by controller vendor. Technically, this conversion can be done using the controller itself in off-line mode, as illustrated in Figure E.11.

Table E.5 Assembly-Level Program for Interrupt

Label	Instruction		Comments
Program area			

Main program

Label	Instruction		Comments	
	AAA	XXX	Instruction n-2	
	BBB	YYY	Instruction n-1	Saves interrupted
Interrupt → CCC		ZZZ	Instruction n ———	program status in stack
	DDD	PPP	Instruction n+1	and jumps to ISR
	EEE	QQQ	Instruction n+2	

Interrupt service routine

ISR	**Programming instructions for finding the source of interrupt**		
	....		
	....		
	JMP	DISR	Jump to digital input service routine

Digital input interrupt service routine

Label	Instruction		Comments
DISR	LDA	DIR0	Load status register DIR0
	ANA	XXX	Extract Bit 7 (diagnostic bit)
	JNZ	WDG	If module is faulty, jump to activate watchdog
	LDA	DOR0	Load status register DOR0
	ANA	XXX	Extract Bit 7 (diagnostic bit)
	JNZ	WDG	If module is faulty, jump to activate watchdog
	LDA	DIR1	Load data-in register DIR1
	STA	IPIB-DIR1	Store in input image buffer IPIB-DIR1
	LDA	IPIB-DIR1	Read input image buffer IPIB-DIR1
	STA	OPIB-DOR1	Store in output image buffer OPIB-DOR1
	LDA	OPIB-DOR1	Load output image buffer OPIB-DOR1
	STA	DOR1	Store in data-out register DOR1
	RTS	**STR**	**Return to main program**
WDG	LDA	YYY	Load mask for extraction of watchdog bit
	STA	WDR1	Drive watchdog
	RTS		**Return** → Restores interrupted program status from stack and returns to interrupted program

Data area		
XXX	'200'	Mask for extraction of diagnostic bit
YYY	'002'	Mask for driving watchdog bit
IPIB-DIR1		Input image
OPIB-DOR1		Output image

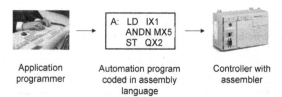

Application   Automation program   Controller with
programmer   coded in assembly   assembler
                     language

Assembler program, residing in Controller, converts the application program
in assembly level language **off-line** into its machine executable equivalent
and stores the same in Controller for its execution **on-line**

Figure E.11 Assembling process.

However, the processor, unlike a general-purpose computer, is optimized for real-time execution of automation functions. It has limited resources for programming environment, so it is not used for the assembling process. This process, done off-line in a **host machine**, is called **cross-assembling**, and it uses **cross-assembler** software supplied by controller vendor.

The host machine is also called the programming device or programming terminal (normally a personal computer or a handheld programmer). Vendor supplied cross-assembler programs, specific to the controller, run on the host machine to produce the machine code specific to the controller. Here, the controller is called the **target machine**. The cross-assembling process is illustrated in Figure E.12.

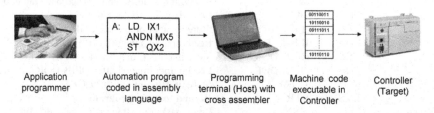

Application   Automation program   Programming   Machine code   Controller
programmer   coded in assembly   terminal (Host) with executable in   (Target)
                  language   cross assembler   Controller

Figure E.12 Cross-assembling and downloading.

The host or the programming terminal, apart from helping the programmer code the program, converts the assembly-level program into a machine executable program and downloads the resultant machine code into the target machine. The host also provides the following additional facilities to the programmer:

- Editing
- Debugging
- Simulation and testing
- Troubleshooting
- Documentation and reporting
- Storing

The handheld programmer, which is computer-based, is compact and more useful as field equipment for troubleshooting at the plant/shop floor.

E.3.7 Higher-Level Programming

Coding the automation program in the assembly level makes the coded automation program executable only on the specific machine/platform for which the program has been coded and not on any other platforms. Hence, programming of the machine is done using the higher-level languages, which are easily understood by the programmers and make the program **portable** for use on any other machines/platforms with a minimum of adaptation.

The programming of the controller for automation functions is done by using the higher-level languages covered by the IEC 61131-3 standard, as discussed in Chapter 10.

Like the assembler program, the compiler program is developed by the platform vendor, and it converts the higher-level language program into its machine executable equivalent and downloads it into the controller (target machine). The processes of compiling, cross-compiling, and downloading are identical to assembling.

Appendix F
Advanced Control Strategies

F.1 Introduction

We discussed the basics of continuous closed loop control and two-step control. The closed loop control has some limitations in its applications. In this appendix, the extensions of the basic closed loop control to overcome these limitations are discussed. Also, extensions of closed loop control for the advanced control strategies, such as feed-forward, cascade, ratio control, are discussed. The chapter concludes with a discussion on multi-step control as an improvement over two-step control.

F.2 Closed Loop Control

In basic continuous closed loop control, the input to the process is continuously controlled to eliminate or minimize the error/deviation between the desired output and the actual output. Figure F.1 illustrates this strategy. In other words, the output is forced to follow the reference input continuously.

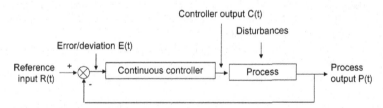

Figure F.1 Continuous control strategy.

Following are the control system parameters, as shown in Figure F.1:

$R(t)$: Reference or set-point input
$P(t)$: Process output
$E(t)$: Error/deviation equal to $R(t) - P(t)$
$C(t)$: Controller output

F.2.1 Controller Response to Control Input

In closed loop continuous control, as discussed in Chapter 6, this is the basic or **proportional control**. This means that the input to the controller is proportional to the difference (deviation) between the actual output and the desired output at any point of time. The typical response of the controller, or the controller output, to a unit step change is shown in Figure F.2.

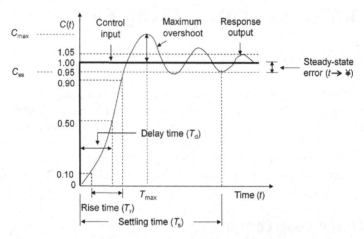

Figure F.2 Controller response to error input.

Following are the terminologies employed in Figure F.2:

- *Maximum overshoot*: $C(t)$ is the response of the controller to the unit step input. C_{max} is the value of the controller output at the maximum overshoot, and C_{ss} is its value at steady state. Maximum overshoot, which usually occurs at the first overshoot, is an indication of the relative stability of the controller. A controller with a large overshoot may be indicating poor stability, and this is undesirable.
- *Delay time* (T_d): Delay time is defined as the time required for the response to reach 50% of its final value.
- *Rise time* (T_r): Rise time is defined as the time required for the response to rise from 10% to 90% of its final value.
- *Steady-state error*: Steady-state error is the desired value within which the controller output finally stays. This is typically within a band of 5% in a good system.
- *Settling time* (T_s): Settling time is defined as the time required for the step response of the controller to reach and stay within the desired value (steady-state value) of the specified band.

As seen in the previous figure, the controller, after receipt of a control input (usually a step input), takes some time to settle down due to its own dynamics. A good controller response should provide the following:

- Less oscillations (more stable)
- Less settling time (faster response)
- Less delay time (faster reaction)
- Less steady-state error (more accurate)

In practice, the traditional proportional control discussed earlier may not meet these listed requirements. In the following sections, different means of achieving these requirements are discussed.

F.2.2 Proportional Control

The basic proportional control (P) is re-illustrated in Figure F.3.

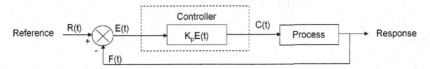

Figure F.3 Proportional control strategy.

The mathematical relationship governing the traditional proportional control strategy is as follows:

$$C(t) = K_p E(t)$$

where

  $C(t)$ is the controller output.
  $R(t)$ is the reference (desired) input.
  $F(t)$ is the feedback or process output.
  $E(t)$ is the error/deviation equal to $R(t) - F(t)$.
  K_p is the proportional gain between the error and the controller output.

Every incremental change in error $E(t)$ produces a corresponding incremental correction (amplification or attenuation) by proportional gain K_p. Pure proportional control reacts to the current error and does not settle at its target value. It retains a residual steady-state error (a function of the proportional gain and the process gain). Despite the steady-state error, it is the proportional control that contributes to the bulk of the output change.

A high proportional gain results in a large change in the response for a given change in the error. If the proportional gain is too high, even though it reduces the residual error, the system can become oscillatory. A small gain results in a small change in the response, even to a large input error. If the proportional gain is too low, the control action may be too small to respond to the disturbances.

F.2.3 Proportional and Integral Control

The problem of steady-state error is overcome in proportional and integral control (PI). The schematic of this control is illustrated in Figure F.4.

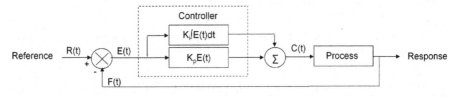

Figure F.4 PI control strategy.

The mathematical relationship governing the proportional and integral control strategy is as follows:

$$C(t) = K_p E(t) + K_i \int E(t)\, dt$$

where

$C(t)$ is the controller output.
$R(t)$ is the reference (desired) input.
$F(t)$ is the feedback or process output.
$E(t)$ is the error/deviation equal to $R(t) - F(t)$.
K_p is the proportional gain between the error and the controller output.
K_i is the integral gain between the controller input and controller output.

The integral control, when added to the proportional control, accelerates the movement of the process toward the reference and eliminates the residual/steady-state error that occurs with a proportional-only controller. Integral control reacts to the sum of recent errors. However, since the control is responding to the accumulated errors from the recent past, it can cause the present value to overshoot the reference value, creating a deviation in the other direction leading to oscillation or instability.

When the residual error is zero, the controller output is fixed at the residual error (as seen in proportion-only control). If the error is not zero, the proportional term contributes a correction, and the integral term begins to increase or decrease the accumulated value, depending on the sign of the error.

F.2.4   Proportional, Integral, and Derivative Control

A combination of proportional, integral, and derivative control (PID) takes care of the problems associated with P and PI, and it produces the best results. The schematic of the proportional, integral, and derivative control is seen in Figure F.5.

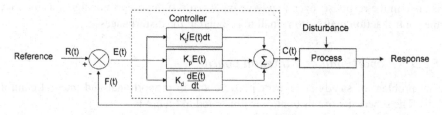

Figure F.5 PID control strategy.

The mathematical relationship governing the proportional, integral, and derivative control strategy is

$$C(t) = K_p e E(t) + K_i \int E(t)\, dt + K_t \frac{dE(t)}{dt}$$

where

C(t) is the controller output.
R(t) is reference (desired) input.
F(t) is feedback or process output.
E(t) is error/deviation equal to R(t) − F(t).
K_p is proportional gain.
K_i is integral gain.
K_d is derivative gain.

The derivative control is the reaction to the rate at which the error has been changing, and it slows the rate of change of the controller output. Hence, the derivative control is used to reduce the magnitude of the overshoot produced by the integral component and to improve the process stability. This is also referred to as rate or anticipatory control. *This mode cannot be used alone because when the error is zero or constant, it does not provide any output.* For every incremental rate of change, this control provides a corresponding incremental change in output by K_d.

In case of a process disturbance, P-only or PI actions cannot react fast enough in returning the process back to set-point without overshoot. Other characteristics are as follows:

- If the process measurement is noisy, the derivative term changes widely and amplifies the noise unless the measurement is filtered.
- The larger the derivative term, the more rapid the controller response to changes in the process value.
- The derivative term reduces both the overshoot and the settling time.

Figure F.6 illustrates the controller response to all three types of inputs namely, proportional, proportional with integral, and proportional with integral and derivative.

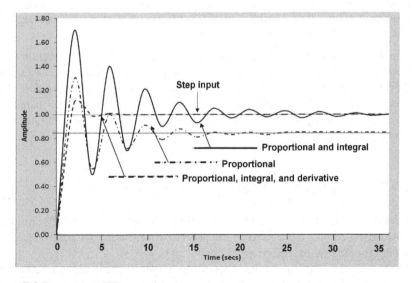

Figure F.6 Response to PID control.

Figure F.7 illustrates an industry application of PID control (speed controller in automobile).

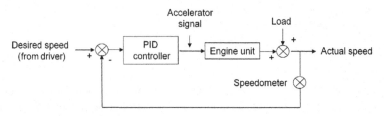

Figure F.7 Example of PID control.

F.2.5 Summary of Control Schemes

The following is the summary of the discussions of various types of closed loop control schemes for continuous process automation:

- *Proportional control (P)*: Basic and essential, but it introduces steady-state error, speeds up the process response and reduces the offset with higher gain. Larger proportional gain (K_p), typically means faster response and larger proportional compensation. An excessively large proportional gain leads to oscillations and instability in the process.
- *Proportional and integral control (PI)*: Eliminates steady-state error, but it introduces oscillations. Larger integral gain (K_i) typically implies quicker elimination of steady-state errors. The trade-off is a large overshoot and oscillations.
- *Proportional, derivative, and integral control (PID)*: Reduces oscillations. Larger derivative gain (K_d) typically decreases the overshoot, but it slows down transient response. However, this may amplify the noise in its response because noise has high-frequency components. It provides an optimum response with proper selection of proportional, integral, and derivative gains.

The selection of what controller modes to use in a process is a function of the process characteristics. Process control loop only regulates the dynamic variables in the process. Controlling parameter is changed to minimize the deviation of the controlled variable. Tuning a control loop is the adjustment of its control parameters (proportional gain/proportional band, integral gain/reset, derivative gain/rate) to the optimum values for the desired controller response.

F.3   Feed-Forward Control

PID controllers act on the error (difference between the set-point and the response) to produce the control input to force the process to follow the set-point while compensating for the external disturbance.

Feed-forward control makes changes to the inputs to the process to counter or pre-empt the anticipated effects of the disturbance in advance. These changes are based on prior knowledge of the effects of the disturbance before it is observed in the process output. Feedback control, when clubbed with the feed-forward control, significantly improves performance of the PID control and, ideally, can entirely eliminate the effect of the measured disturbance on the process output. To implement the feed-forward control, the time taken for the disturbance to affect the output should not be longer than the time taken for the feed-forward controller to affect the output. Figure F.8 illustrates the schematic of feed-forward control.

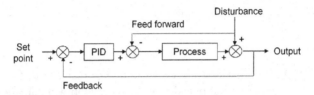

Figure F.8 Feed-forward control.

Figure F.9 illustrates the industry application of feed-forward control of a two-element boiler. In the boiler, the two variables (steam flow and drum water level) influence the feed water valve position. In the normal situation, the PID controller regulates the drum water level based on its set-point without taking into consideration the effect of steam flow on the drum water level. In the feed-forward case, the PID controller output is combined with the steam flow to compensate the effect of steam flow on the drum water level. In other words, the feed-forward control gives a better regulation of the feed water to respond to any change in the steam demand.

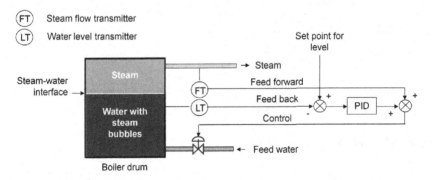

Figure F.9 Feed-forward control—example.

F.4   Cascade Control

A controlled process may, in some cases, be considered as two processes in series or as a cascaded process in which the output of the first process is measured. The output of the master controller is the set-point for the slave controller. This control provides a better dynamic performance. Figure F.10 illustrates the schematic of cascade control.

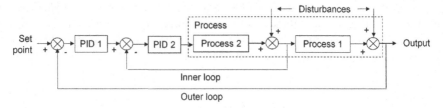

Figure F.10 Cascade control.

The benefit of using the cascade configuration is that any disturbances within the inner loop can be corrected quickly by the outer loop without waiting for them to show up in the outer loop. This gives a better (tighter and faster) overall control. The outer loop can be relatively tuned for a faster response. It responds to the disturbances quickly and minimizes any fluctuations in its output due to disturbance. The outer loop still controls the final output.

Figure F.11 illustrates an industry application of cascade control in heat exchangers. Heat exchangers are used to heat or cool a process fluid to a desired temperature by steam. The output of the master controller (temperature) manipulates the set-point of the slave controller (steam flow). This eliminates the effects of feed or product disturbances to improve the performance of the control loop.

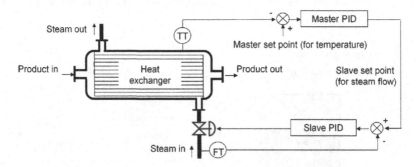

Figure F.11 Cascade control—example.

Combinations of single- and multi-loop controllers can be configured to work as cascade controllers.

F.5 Ratio Control

This is a technique to control a process variable at a set-point, which is calculated as a proportion (ratio) of an uncontrolled or lead input variable. The set-point ratio determines the proportion of the lead value to be used as the actual control set-point. The set-point ratio can be either greater or lesser than the lead input. Figure F.12 illustrates the schematic of ratio control.

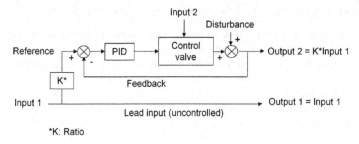

*K: Ratio

Figure F.12 Ratio control.

Ratio control systems are used extensively to optimize the relationship of the mixture between two flows in processing plants. In other words, ratio control is used to ensure that two or more flows are kept at the same ratio even if the lead flow is changing. The typical applications are as follows:

- Blending two or more flows to produce a mixture with specified composition
- Blending two or more flows to produce a mixture with specified physical properties
- Air–fuel ratio control

Figure F.13 illustrates an industry application of ratio control (air–fuel mixer).

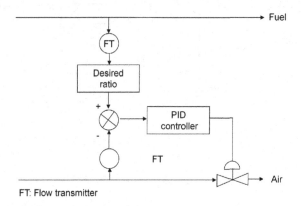

FT: Flow transmitter

Figure F.13 Ratio control—example.

F.6   Multi-Step Control

Multi-step control is similar to two-step control, as discussed in Chapter 6, but with at least one more intermediate step. This intermediate step further reduces the possibility of oscillations or hunting around the set-point. In other words, with more steps, finer or near continuous control is possible. The multi-step control is best explained with the example of a domestic voltage stabilizer with 3-step controller.

The function of the domestic voltage stabilizer is to provide regulated voltage output within the acceptable range of 220 volts ±10%, or 198–242 volts, from an unregulated input voltage. The specifications of the stabilizer are as follows:

- The stabilizer is designed to work within the input range of 176–264 volts, and it gets cut off from the mains if the input is outside that range.
- If the input voltage falls below 198 volts, Switch 1 is closed to output 110% of the input voltage (Step 1). In other words, a +10% correction is applied.
- If the input voltage is within 198–242 volts, Switch 2 is closed to follow the input (Step 2). In other words, no correction is applied.
- If the input voltage rises above 242 volts, Switch 3 is closed to output 90% of the input voltage (Step 3). In other words, a −10% correction is applied.
- To sum up, the output is made to stay within the acceptable range (198–242 volts) provided the input range is between 176 and 264 volts.

Figure F.14 illustrates the control schematic of the voltage stabilizer with three steps.

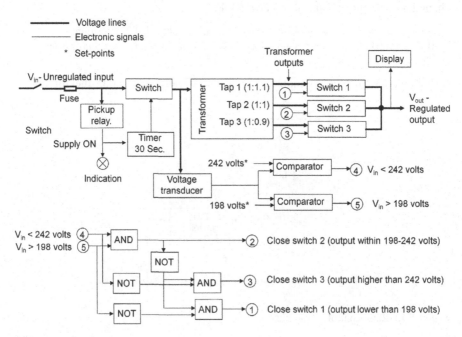

Figure F.14 3-Step control.

Figure F.15 illustrates an industry example of a voltage stabilizer with 5-step control, supply indication, and input/output voltage display. Other specifications are as follows:

- *Input voltage range*: 100–290V AC.
- *Output voltage range*: 200–250V AC (from 100V to 280V input).
- *Low voltage cutoff*: 70V input with 138V output.
- *High-voltage cutoff*: 300V input with 269V output.

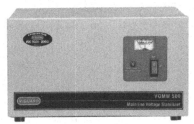

Courtesy: www.vguard.in

Figure F.15 Voltage stabilizer with 5-step control.

Appendix G
Power Supply System

G.1   Introduction

The following information highlights the importance of a reliable power source for the automation system.

All the subsystems of the automation system (instrumentation, control, and human interface) should be available as long as the process is in operation and producing the result. In Chapter 14, we discussed the need to provide standby or redundancy to vital components so that the failure of any vital component does not affect the functioning of the automation system. However, any failure of the external power supply leads to the total unavailability of the automation system even though the latter may be healthy. This chapter discusses the solutions to overcome this problem.

G.2   Float-cum-Boost Charger with Battery Backup

Normally, float-cum-boost chargers with battery backups are employed where controllers are installed in remote places and requiring no-break DC power. Figure G.1

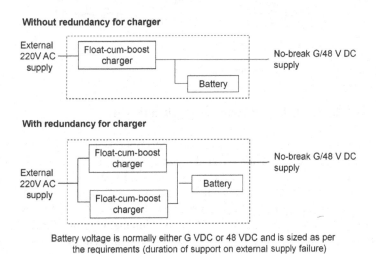

Figure G.1 Charger configurations.

illustrates the two configurations for this arrangement. In the configuration without redundancy for the charger, during normal operation (when the external supply is present), the charger charges the battery (trickle charging) and feeds the controller simultaneously. In other words, during normal operation, the battery floats. When the external input power supply fails, the battery takes over to feed the controller (bumpless switchover).

In the second configuration (with redundancy for the charger), the charger is provided with a standby. Each charger is individually equipped to take the full load, and both the chargers share the total load (trickle charging and feeding controller). Trickle charging compensates for the loss of power in the battery while in floating mode. Boost charging charges the battery in off-line mode when the battery is fully discharged. When any one of the chargers fails, the healthy charger takes the full load. This arrangement provides time to get the faulty charger repaired and re-installed.

If necessary, there can be a standby for battery also.

Figure G.2 illustrates an industry example of a charger.

Non-redundant charger   Redundant charger

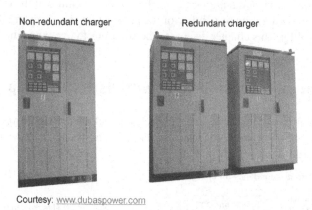

Courtesy: www.dubaspower.com

Figure G.2 Nonredundant and redundant chargers.

G.3   Uninterrupted Power Supply System

The uninterrupted power supply system (UPSS) approach is employed where the automation equipment is installed in the control center (controllers, operator stations, servers, etc.), requiring no-break AC power. Figure G.3 illustrates the two configurations for this arrangement. In the configuration without redundancy for the rectifier/inverter, during normal operation (when the external supply is present), the rectifier simultaneously charges the battery (trickle charging) and feeds the inverter. The inverter, in turn, converts the DC power to AC power and feeds the automation equipment. In other words, during normal operation, the battery floats. When the external power supply fails, the battery takes over to feed the inverter which then feeds the automation equipment (bumpless switchover).

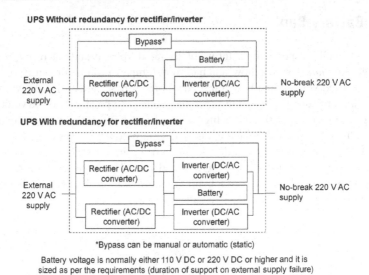

Figure G.3 UPSS configurations.

In the second configuration (with redundancy for charger/inverter with each rectifier/inverter individually equipped to take the full load), both the rectifiers/inverters share the total load (trickle charging and feeding the automation equipment). When any one of the rectifiers or inverters fails, the healthy rectifier or inverter takes the full load.

Here also, there can be a standby battery if the situation warrants. Additional availability can be provided with a bypass to switch the external power supply to the automation equipment when the entire UPSS fails. Bypass is a switch to connect to the output mains if the UPSS fails, and it can be either manual or automatic (static).

Figure G.4 illustrates industry examples of UPSS.

Figure G.4 Nonredundant and redundant UPSS.

G.4   Battery Bank

The battery cell is specified in terms of its rated voltage and rated ampere-hours (Ahs). *Ampere hours* means how much current (ampere) can be drawn from the battery at the rated voltage and how long. The more current (ampere) is drawn, the less is its duration of support. To get the specified voltage (typically 24 or 48 volts in case of charger and 110 or 220 volts or higher in case of UPSS), the required number of battery cells of are stacked up and connected serially to form a battery bank or stack. Normally, the battery floats and draws power from the charger/rectifier only for its trickle charging to compensate for its no-load loss. The batteries need to be boost charged after sustained use to bring them to a normal level. Figure G.5 illustrates an example of an industrial battery and a battery bank.

Cell   Bank
Courtesy: www.exide.co.in

Figure G.5 Battery cell and battery bank.

G.5   Power Distribution

For a good installation, it is not enough if we have a charger or UPSS for power backup. Apart from the automation equipment, there may be other equipment that

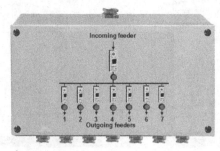

Courtesy: www.fabionix.co.in

Figure G.6 Power distribution panel.

needs the no-break power. To facilitate this and to provide proper distribution of the load for all the equipment that needs no-break power, a distribution panel/board is normally employed, as illustrated in Figure G.6. As seen here, the incoming feeder (no-break supply from the charger or UPSS) and the outgoing feeders to all the equipment (including automation equipment) are provided with adequate protection and display facilities (MCB and indication lamp).

Appendix H
Further Reading

This book is mainly based on the author's experience learning, practicing, and teaching automation in ABB, and no specific references are followed. The best material for further reading is the information published by the automation companies on their products, systems, and solutions. In addition to this, the following are some good references for further reading:

General

Dimon, T.G., 2003. The automation, systems, and instrumentation dictionary, 4th ed. Instrument Society of America, Research Triangle Park, N.C.

Herb, S.M., 2004. Understanding distributed processor systems for control. Instrument Society of America, Research Triangle Park, N.C.

Liptak, B.G. (Ed.), 1995. Instrument engineers' handbook, Vol. 1: Process measurement and analysis. Butterworth Heinmann Ltd., Oxford.

Liptak, B.G. (Ed.), 1995. Instrument engineers' handbook, Vol. 2: Process control. Butterworth Heinmann Ltd., Oxford.

Liptak, B.G. (Ed.), 2002. Instrument engineers' handbook, Vol. 3: Process software and digital networks. CRC Press, Boca Raton, London.

Software

Kamal, R., 2003. Embedded systems: architecture, programming and design. Tata McGraw Hill, New Delhi.

Programming

SIMATIC Ladder Logic (LAD) for S7-300 and S7-400 programming reference manual. This manual is part of the documentation package with the order number 6ES7810-4CA08-8BW1, Edition 03/2006 A5E00706949-01.

SIMATIC Function Block Diagram (FBD) for S7-300 and S7-400 programming reference manual. This manual is part of the documentation package with the order number 6ES7810-4CA07-8BW1, Edition 01/2004 A5E00261409-01.

Instrumentation

Whitt, M.D., 2004. Successful instrumentation and control systems design. Instrument Society of America, Research Triangle Park, N.C.

Fieldbus

Berge, J., 2004. Fieldbuses for process control: engineering, operation and maintenance. Instrument Society of America, Research Triangle Park, N.C.

Chapman, P.W., 1996. Smart sensors. Instrument Society of America, Research Triangle Park, N.C.

Mitchell, R., 2004. PROFIBUS: a pocket guide. Instrument Society of America, Research Triangle Park, N.C.

Verhappen, I., 2004. Foundation fieldbus: a pocket guide. Instrument Society of America, Research Triangle Park, N.C.

Programmable Logic Controller

Hughes, T.A., 2005. Programmable controllers, 4th ed. Instrument Society of America, Research Triangle Park, N.C.

PC-Based Controller and Programmable Automation Controller

http://www.opto22.com
http://www.advantech.com

Industrial Control

Astrom, K.J., 1995. PID controllers: theory, design and tuning. Instrument Society of America, Research Triangle Park, N.C.

Coggan, D.A. (Ed.), 2005. Fundamentals of industrial control (2nd ed.). Instrument Society of America, Research Triangle Park, N.C.

Corripio, A.B., 1998. Design and application of process control systems. Instrument Society of America, Research Triangle Park, N.C.

Corripio, A.B., 2001. Tuning of industrial control systems, 2nd ed. Instrument Society of America, Research Triangle Park, N.C.

Johnson, C.D., 2005. Process control instrumentation technology, 7th ed. Pearson Education, New Delhi.

McMillan, G.K., 2000. Good tuning: a pocket guide. Instrument Society of America, Research Triangle Park, N.C.

Communication and Networking

Caro, D., 2004. Automation network selection. Instrument Society of America, Research Triangle Park, N.C.

Caro, D., 2004. Wireless networks for industrial automation. Instrument Society of America, Research Triangle Park, N.C.

Marshall, P.S., 2004. Industrial Ethernet: how to plan, install, and maintain TCP/IP Ethernet networks: the basic reference guide for automation and process control engineers, 2nd ed. Instrument Society of America, Research Triangle Park, N.C.

Thompson, L.M., 2002. Industrial data communications, 3rd ed. Instrument Society of America, Research Triangle Park, N.C.

SCADA

Boyer, S.A., 2004. SCADA: supervisory control and data acquisition, 3rd ed. Instrument Society of America, Research Triangle Park, N.C.

Others

Nisenfeld, A.E. (Ed.), 1996. Batch control: practical guides for measurement and control. Instrument Society of America, Research Triangle Park, N.C.

Park, J., 2003. Practical data acquisition for instrumentation and control systems. Newnes (Imprint of Elsevier), Research Triangle Park, N.C.

Platt, G., 1998. Process control: a primer for the nonspecialist and the newcomer, 2nd ed. Instrument Society of America, Research Triangle Park, N.C.

Rossiter, J.A., 2003. Model-based predictive control: a practical approach. CRC Press, Boca Raton; London.

Printed in the United States
By Bookmasters